INTRODUCTION TO COSMOLOGY

INTRODUCTION TO COSMOLOGY

Barbara Ryden
The Ohio State University

San Francisco Boston New York
Capetown Hong Kong London Madrid Mexico City
Montreal Munich Paris Singapore Sydney Tokyo Toronto

Acquisitions Editor: Adam Black
Project Editor: Nancy Benton
Production Editor: Joan Marsh
Text Designer: Leslie Galen
Cover Designers: Blakeley Kim, Kenneth Probst
Marketing Manager: Christy Lawrence
Manufacturing Coordinator: Vivian McDougal
Project Coordination and Electronic Page Makeup: Integre Technical Publishing Co., Inc.

Library of Congress Cataloging-in-Publication Data
Ryden, Barbara Sue.
Introduction to cosmology : Barbara Ryden.
p. cm.
Includes bibliographical references and index.
ISBN 0-8053-8912-1 (hardcover)
1. Cosmology. I. Title.
QB981 .R93 2003
523.1—dc21 2002013176

ISBN 0-8053-8912-1

2 3 4 5 6 7 8 9 10 —MAL— 05 04 03
www.aw.com/aw

Preface

This book is based on my lecture notes for an upper-level undergraduate cosmology course at The Ohio State University. The students taking the course were primarily juniors and seniors majoring in physics and astronomy. In my lectures, I assumed that my students, having triumphantly survived freshman and sophomore physics, had a basic understanding of electrodynamics, statistical mechanics, classical dynamics, and quantum physics. As far as mathematics was concerned, I assumed that, like modern major generals, they were very good at integral and differential calculus. Readers of this book are assumed to have a similar background in physics and mathematics. In particular, no prior knowledge of general relativity is assumed; the (relatively) small amounts of general relativity needed to understand basic cosmology are introduced as needed.

Unfortunately, the National Bureau of Standards has not gotten around to establishing a standard notation for cosmological equations. It seems that every cosmology book has its own notation; this book is no exception. My main motivation was to make the notation as clear as possible for the cosmological novice.

I hope that reading this book will inspire students to further explorations in cosmology. The annotated bibliography at the end of the text provides a selection of recommended cosmology books, at the popular, intermediate, and advanced levels.

Many of the illustrations in this book were adapted from figures in published scientific papers. My thanks go to the authors of those papers for granting permission to use their work. Particular thanks are due to Avishai Dekel (Fig. 2.2), Wendy Freedman (Fig. 2.5), Adam Riess (Figs. 7.5 and 7.6), Ray White (Figs 8.1 and 8.2), Sidney van den Bergh (Fig. 8.4), Warrick Couch (Fig. 8.7), David Leisawitz (Figs. 9.1 and 9.2), John Ruhl (Fig. 9.5), Xiaomin Wang (Fig. 9.6), Paul Richards (Fig. 9.7), Davis Tytler (Fig. 10.4), Scott Burles (Fig. 10.5), Matthew Colless (Fig. 12.1), and Patrick Hayes (Fig. 12.2).

Many people (too many to name individually) helped in the making of this book; I thank them all. I owe particular thanks to the students who took my undergraduate cosmology course at Ohio State University. Their feedback (including nonverbal feedback such as frowns and snores during lectures) greatly improved the lecture notes on which this book is based. Adam Black and Nancy Benton at Addison Wesley made possible the great leap from rough lecture notes to polished book. The reviewers of the text, both anonymous and onymous, pointed out many errors and omissions. I owe particular thanks to Gerald Newsom and Joel Primack, whose careful reading of the manuscript improved it greatly. My greatest debt, however, is to Rick Pogge, who acted as my computer maven, graphics guru, and sanity check. (He was also a tireless hunter of creeping fox terrier clones.) As a small sign of my great gratitude, this book is dedicated to him.

LIST OF REVIEWERS

Peter Becker, George Mason University
Ed Bertschinger, Massachusetts Institute of Technology
Sukanta Bose, Washington State University
Ed Churchwell, University of Wisconsin at Madison
Marc Davis, University of California at Berkeley
Nick Gnedin, University of Colorado
Larry Marschall, Gettysburg College
Hugo Martel, University of Texas at Austin
Adrian Melott, University of Kansas
Charles Niederriter, Gustavus Adolphus College
T. Padmanabhan, Inter-University Centre for Astronomy and Astrophysics
Joel Primack, University of California at Santa Cruz
Donald Salisbury, Austin College
James Schombert, University of Oregon
Clarke Wellborn, Brevard College
Edward Wright, University of California at Los Angeles
Jose Wudka, University of California at Riverside
Nicolle Zellner, Rensselaer Polytechnic Institute

Contents

CHAPTER 1

Introduction

Cosmology is the study of the universe, or cosmos, regarded as a whole. Attempting to cover the study of the entire universe in a single volume may seem like a megalomaniac's dream. The universe, after all, is richly textured, with structures on a vast range of scales; planets orbit stars, stars are collected into galaxies, galaxies are gravitationally bound into clusters, and even clusters of galaxies are found within larger superclusters. Given the complexity of the universe, the only way to condense its history into a single book is by a process of ruthless simplification. For much of this book, therefore, we will be considering the properties of an idealized, perfectly smooth, model universe. Only near the end of the book will we consider how relatively small objects, such as galaxies, clusters, and superclusters, are formed as the universe evolves. It is amusing to note in this context that the words *cosmology* and *cosmetology* come from the same Greek root: the word *kosmos*, meaning harmony or order. Just as cosmetologists try to make a human face more harmonious by smoothing over small blemishes such as pimples and wrinkles, cosmologists sometimes must smooth over small "blemishes" such as galaxies.

A science that regards entire galaxies as being small objects might seem, at first glance, very remote from the concerns of humanity. Nevertheless, cosmology deals with questions that are fundamental to the human condition. The questions that vex humanity are given in the title of a painting by Paul Gauguin (Figure 1.1): "Where do we come from? What are we? Where are we going?" Cosmology grapples with these questions by describing the past, explaining the present, and predicting the future of the universe. Cosmologists ask questions such as "What is the universe made of? Is it finite or infinite in spatial extent? Did it have a beginning some time in the past? Will it come to an end some time in the future?"

Cosmology deals with distances that are very large, objects that are very big, and timescales that are very long. Cosmologists frequently find that the standard SI units are not convenient for their purposes: the meter (m) is awkwardly short, the kilogram (kg) is awkwardly tiny, and the second (s) is awkwardly brief. Fortunately, we can adopt the units that have been developed by astronomers for dealing with large distances, masses, and times.

One distance unit used by astronomers is the astronomical unit (AU), equal to the mean distance between the Earth and Sun; in metric units, $1\,\text{AU} = 1.5 \times 10^{11}\,\text{m}$. Although the astronomical unit is a useful length scale within the solar

FIGURE 1.1 *Where Do We Come From? What Are We? Where Are We Going?* Paul Gauguin, 1897. [Museum of Fine Arts, Boston]

system, it is small compared to the distances between stars. To measure interstellar distances, it is useful to use the parsec (pc), equal to the distance at which 1 AU subtends an angle of 1 arcsecond; in metric units, $1\,\text{pc} = 3.1 \times 10^{16}\,\text{m}$. For example, we are at a distance of 1.3 pc from Proxima Centauri (the Sun's nearest neighbor among the stars) and 8500 pc from the center of our galaxy. Although the parsec is a useful length scale within our galaxy, it is small compared to the distances between galaxies. To measure intergalactic distances, we use the megaparsec (Mpc), equal to 10^6 pc, or 3.1×10^{22} m. For example, we are at a distance of 0.7 Mpc from M31 (otherwise known as the Andromeda galaxy) and 15 Mpc from the Virgo cluster (the nearest big cluster of galaxies).

The standard unit of mass used by astronomers is the solar mass ($\text{M}_\odot$); in metric units, the Sun's mass is $1\,\text{M}_\odot = 2.0 \times 10^{30}\,\text{kg}$. The total mass of our galaxy is not known as accurately as the mass of the Sun; in round numbers, though, it is $M_{\text{gal}} \approx 10^{12}\,\text{M}_\odot$. The Sun, incidentally, also provides the standard unit of power used in astronomy. The Sun's luminosity (that is, the rate at which it radiates away energy in the form of light) is $1\,\text{L}_\odot = 3.8 \times 10^{26}$ watts. The total luminosity of our galaxy is $L_{\text{gal}} = 3.6 \times 10^{10}\,\text{L}_\odot$.

For times much longer than a second, astronomers use the year (yr), defined as the time it takes the Earth to go once around the Sun. One year is approximately equal to 3.2×10^7 s. In a cosmological context, a year is frequently an inconveniently short period of time, so cosmologists often use gigayears (Gyr), equal to 10^9 yr, or 3.2×10^{16} s. For example, the age of the Earth is more conveniently written as 4.6 Gyr than as 1.5×10^{17} s.

In addition to dealing with very large things, cosmology also deals with very small things. Early in its history, as we shall see, the universe was very hot and dense, and some interesting particle physics phenomena were occurring. Consequently, particle physicists have plunged into cosmology, introducing some terminology and units of their own. For instance, particle physicists tend to measure energy units in electron volts (eV) instead of joules (J). The conversion factor between electron volts and joules is $1\,\text{eV} = 1.6 \times 10^{-19}\,\text{J}$. The rest energy of an electron, for instance, is $m_e c^2 = 511{,}000\,\text{eV} = 0.511\,\text{MeV}$, and the rest energy of a proton is $m_p c^2 = 938.3\,\text{MeV}$.

When you stop to think of it, you realize that the units of meters, megaparsecs, kilograms, solar masses, seconds, and gigayears could only be devised by ten-fingered Earthlings obsessed with the properties of water. An eighteen-tentacled silicon-based lifeform from a planet orbiting Betelgeuse would probably devise a different set of units. A more universal, less culturally biased system of units is the Planck system, based on the universal constants G, c, and $\hbar$. Combining the Newtonian gravitational constant, $G = 6.7 \times 10^{-11}\,\text{m}^3\,\text{kg}^{-1}\,\text{s}^{-2}$, the speed of light, $c = 3.0 \times 10^8\,\text{m}\,\text{s}^{-1}$, and the reduced Planck constant, $\hbar = h/(2\pi) = 1.1 \times 10^{-34}\,\text{J}\,\text{s} = 6.6 \times 10^{-16}\,\text{eV}\,\text{s}$, yields a unique length scale, known as the Planck length:

$$\ell_P \equiv \left(\frac{G\hbar}{c^3}\right)^{1/2} = 1.6 \times 10^{-35}\,\text{m}. \tag{1.1}$$

The same constants can be combined to yield the Planck mass,[1]

$$M_P \equiv \left(\frac{\hbar c}{G}\right)^{1/2} = 2.2 \times 10^{-8}\,\text{kg}, \tag{1.2}$$

and the Planck time,

$$t_P \equiv \left(\frac{G\hbar}{c^5}\right)^{1/2} = 5.4 \times 10^{-44}\,\text{s}. \tag{1.3}$$

Using Einstein's relation between mass and energy, we can also define the Planck energy,

$$E_P = M_P c^2 = 2.0 \times 10^9\,\text{J} = 1.2 \times 10^{28}\,\text{eV}. \tag{1.4}$$

By bringing the Boltzmann constant, $k = 8.6 \times 10^{-5}\,\text{eV}\,\text{K}^{-1}$, into the act, we can also define the Planck temperature,

$$T_P = E_P/k = 1.4 \times 10^{32}\,\text{K}. \tag{1.5}$$

When distance, mass, time, and temperature are measured in the appropriate Planck units, then $c = k = \hbar = G = 1$. This is convenient for individuals who have difficulty in remembering the numerical values of physical constants. However, using Planck units can have potentially confusing side effects. For instance, many cosmology texts, after noting that $c = k = \hbar = G = 1$ when Planck units are used, then proceed to omit c, k, $\hbar$, and/or G from all equations. For instance, Einstein's celebrated equation, $E = mc^2$, becomes $E = m$. The blatant dimensional incorrectness of such an equation is jarring, but it simply means that

[1] The Planck mass is roughly equal to the mass of a grain of sand a quarter of a millimeter across.

the rest energy of an object, measured in units of the Planck energy, is equal to its mass, measured in units of the Planck mass. In this book, however, I will retain all factors of c, k, $\hbar$, and G, for the sake of clarity.

Here we will deal with distances ranging from the Planck length to 10^4 Mpc or so, a span of some 61 orders of magnitude. Dealing with such a wide range of length scales requires a stretch of the imagination, to be sure. However, cosmologists are not permitted to let their imaginations run totally unfettered. Cosmology, I emphasize strongly, is based ultimately on observation of the universe around us. Even in ancient times, cosmology was based on observations; unfortunately, those observations were frequently imperfect and incomplete. Ancient Egyptians, for instance, looked at the desert plains stretching away from the Nile valley and the blue sky overhead. Based on their observations, they developed a model of the universe in which a flat Earth (symbolized by the earth god Geb in Figure 1.2) was covered by a solid dome (symbolized by the sky goddess Nut). Greek cosmology was based on more precise and sophisticated observations. Ancient Greek astronomers deduced, from their observations, that the Earth and Moon are spherical, that the Sun is much farther from the Earth than the Moon is, and that the distance from the Earth to the stars is much greater than the Earth's diameter. Based on this knowledge, Greek cosmologists devised a "two-sphere" model of the universe, in which the spherical Earth is surrounded by a much larger celestial sphere, a spherical shell to which the stars are attached. Between the Earth and the celestial sphere, in this model, the Sun, Moon, and planets move on their complicated apparatus of epicycles and deferents.

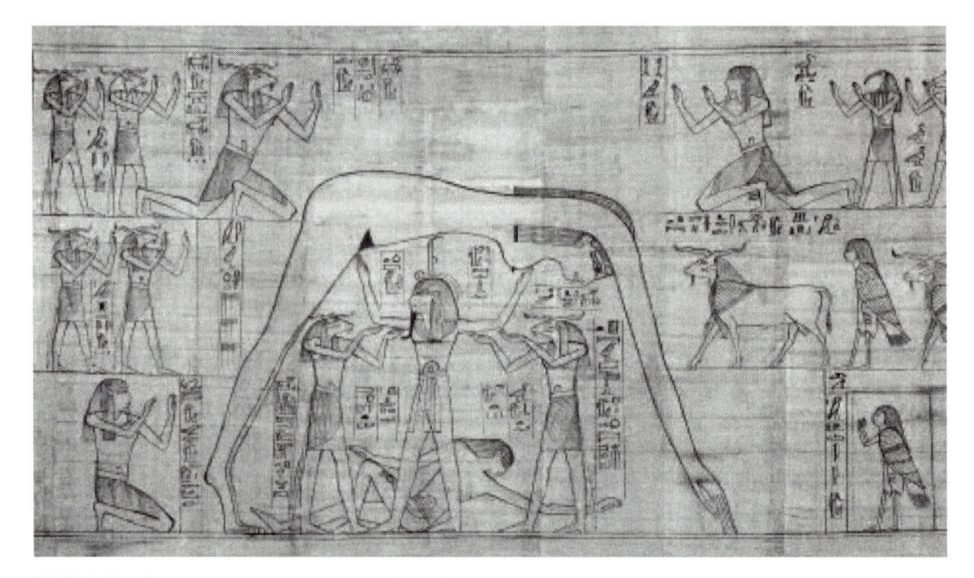

FIGURE 1.2 The ancient Egyptian view of the cosmos: the sky goddess Nut, supported by the air god Shu, arches over the earth god Geb (from the Greenfield Papyrus, ca. 1025 BC). [©Copyright The British Museum]

Although cosmology is ultimately based on observation, sometimes observations temporarily lag behind theory. During periods when data are lacking, cosmologists may adopt a new model for aesthetic or philosophical reasons. For instance, when Copernicus proposed a new Sun-centered model of the universe, to replace the Earth-centered two-sphere model of the Greeks, he didn't base his model on new observational discoveries. Rather, he believed that putting the Earth in motion around the Sun resulted in a conceptually simpler, more appealing model of the universe. Direct observational evidence didn't reveal that the Earth revolves around the Sun, rather than vice versa, until the discovery of the aberration of starlight in the year 1728, nearly two centuries after the death of Copernicus. Foucault didn't demonstrate the rotation of the Earth, another prediction of the Copernican model, until 1851, over *three* centuries after the death of Copernicus. However, although observations sometimes lag behind theory in this way, every cosmological model that isn't eventually supported by observational evidence must remain pure speculation.

The current standard model for the universe is the "Hot Big Bang" model, which states that the universe has expanded from an initially hot and dense state to its current relatively cool and tenuous state, and that the expansion is still going on today. To see why cosmologists have embraced the Hot Big Bang model, let us turn, in the next chapter, to the fundamental observations on which modern cosmology is based.

SUGGESTED READING

Full references are given in the Annotated Bibliography on page 235.

Cox (2000): Accurate values of physical and astronomical constants

Harrison (2000), ch. 1–4: A history of early (pre-Einstein) cosmology

CHAPTER

2 Fundamental Observations

Some of the observations on which modern cosmology is based are highly complex, requiring elaborate apparatus and sophisticated data analysis. However, other observations are surprisingly simple. Let's start with an observation that is deceptive in its extreme simplicity.

2.1 ■ THE NIGHT SKY IS DARK

Step outside on a clear, moonless night, far from city lights, and look upward. You will see a dark sky, with roughly two thousand stars scattered across it. The fact that the night sky is dark at visible wavelengths, instead of being uniformly bright with starlight, is known as *Olbers' Paradox*, after the astronomer Heinrich Olbers, who wrote a scientific paper on the subject in 1826. As it happens, Olbers was not the first person to think about Olbers' Paradox. As early as 1576, Thomas Digges mentioned how strange it is that the night sky is dark, with only a few pinpoints of light to mark the location of stars.[1]

Why should it be paradoxical that the night sky is dark? Most of us simply take for granted the fact that daytime is bright and nighttime is dark. The darkness of the night sky certainly posed no problems to the ancient Egyptians or Greeks, to whom stars were points of light stuck to a dome or sphere. However, the cosmological model of Copernicus required that the distance to stars be very much larger than an astronomical unit; otherwise, the parallax of the stars, as the Earth goes around on its orbit, would be large enough to see with the naked eye. Moreover, since the Copernican system no longer requires that the stars be attached to a rotating celestial sphere, the stars can be at different distances from the Sun. These liberating realizations led Thomas Digges, and other post-Copernican astronomers, to embrace a model in which stars are large glowing spheres, like the Sun, scattered throughout infinite space.

Let's compute how bright we expect the night sky to be in an infinite universe. Let n be the average number density of stars in the universe, and let L be the average stellar luminosity. The flux received here at Earth from a star of luminosity

[1]The name "Olbers' Paradox" is thus a prime example of what historians of science jokingly call the law of misonomy: nothing is ever named after the person who really discovers it.

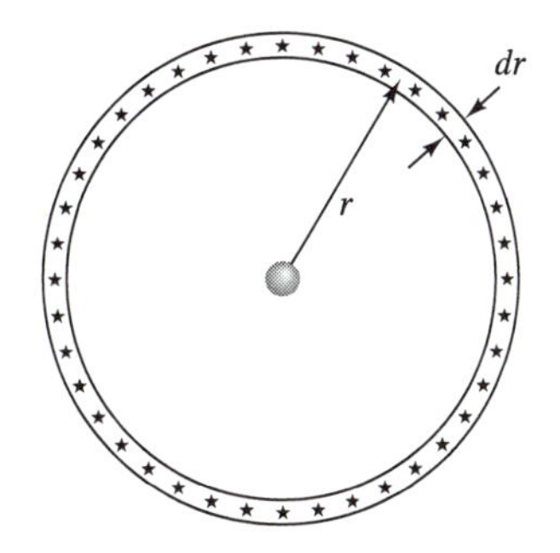

FIGURE 2.1 A star-filled spherical shell, of radius r and thickness dr, centered on the Earth.

L at a distance r is given by an inverse square law:

$$f(r) = \frac{L}{4\pi r^2}. \tag{2.1}$$

Now consider a thin spherical shell of stars, with radius r and thickness dr, centered on the Earth (Figure 2.1). The intensity of radiation from the shell of stars (that is, the power per unit area per steradian of the sky) is

$$dJ(r) = \frac{L}{4\pi r^2} \cdot n \cdot r^2 dr = \frac{nL}{4\pi} dr. \tag{2.2}$$

The total intensity of starlight from a shell thus depends only on its thickness, not on its distance from us. We can compute the total intensity of starlight from *all* the stars in the universe by integrating over shells of all radii:

$$J = \int_{r=0}^{\infty} dJ = \frac{nL}{4\pi} \int_0^{\infty} dr = \infty. \tag{2.3}$$

Thus, I have demonstrated that the night sky is infinitely bright.

This is utter nonsense.

Therefore, one (or more) of the assumptions that went into the above analysis of the sky brightness must be wrong. Let's scrutinize some of the assumptions. One assumption that I made is that we have an unobstructed line of sight to every star in the universe. This is not true. In fact, since stars have a finite angular size as seen from Earth, nearby stars will hide more distant stars from our view. Nevertheless, in an infinite distribution of stars, every line of sight should end at the surface of a star; this would imply a surface brightness for the sky equal to the surface brightness of a typical star. This is an improvement on an infinitely bright sky, but is still distinctly different from the dark sky we actually see. Heinrich Olbers himself tried to resolve Olbers' Paradox by proposing that distant stars are hidden from view by interstellar matter that absorbs starlight. This resolution does not work, because the interstellar matter would be heated by starlight until it had the same temperature as the surface of a star. At that point, the interstellar matter would emit as much light as it absorbs, and glow as brightly as the stars themselves.

A second assumption I made is that the number density n and mean luminosity L of stars are constant throughout the universe; more accurately, the assumption made in equation (2.3) is that the product nL is constant as a function of r. This might not be true. Distant stars might be less luminous or less numerous than nearby stars. If we are in a clump of stars of finite size, then the absence of stars at large distances will keep the night sky from being bright. Similarly, if distant stars are sufficiently low in luminosity compared to nearby stars, then they won't contribute significantly to the sky brightness. In order for the integrated intensity in equation (2.3) to be finite, the product nL must fall off more rapidly than $nL \propto 1/r$ as $r \to \infty$.

A third assumption is that the universe is infinitely large. This might not be true. If the universe extends only to a maximum distance r_{max} from us, then the total intensity of starlight we see in the night sky will be $J \sim nLr_{max}/(4\pi)$. Note that this result will also be found if the universe is infinite in space, but is devoid of stars beyond a distance r_{max}.

A fourth assumption, slightly more subtle than the previous ones, is that the universe is infinitely old. This might not be true. Because the speed of light is finite, when we look farther out in space, we are looking farther out in time. Thus, we see the Sun as it was 8.3 minutes ago, Proxima Centauri as it was 4 years ago, and M31 as it was 2 million years ago. If the universe has a finite age t_0, the intensity of starlight we see at night will be at most $J \sim nLct_0/(4\pi)$. Note that this result will also be found if the universe is infinitely old, but has only contained stars for a finite time t_0.

A fifth assumption is that the flux of light from a distant source is given by the inverse square law of equation (2.1). This might not be true. The assumption that $f \propto 1/r^2$ would have seemed totally innocuous to Olbers and other nineteenth-century astronomers; after all, the inverse square law follows directly from Euclid's laws of geometry. However, in the twentieth century, Albert Einstein, that great questioner of assumptions, demonstrated that the universe might not obey the laws of Euclidean geometry. In addition, the inverse square law assumes that the source of light is stationary relative to the observer. If the universe is systematically expanding or contracting, then the light from distant sources will be redshifted to lower photon energies or blueshifted to higher photon energies.

Thus, the infinitely large, eternally old, Euclidean universe that Thomas Digges and his successors pictured simply does not hold up to scrutiny. This is a textbook, not a suspense novel, so I'll tell you right now: the primary resolution to Olbers' Paradox comes from the fact that the universe has a finite age. The stars beyond some finite distance, called the horizon distance, are invisible to us because their light hasn't had time to reach us yet. A particularly amusing bit of cosmological trivia is that the first person to hint at the correct resolution of Olbers' Paradox was Edgar Allen Poe.[2] In his essay "Eureka: A Prose Poem," completed in 1848,

[2] That's right, the "Nevermore" guy.

Poe wrote, "Were the succession of stars endless, then the background of the sky would present us an [*sic*] uniform density... since there could be absolutely no point, in all that background, at which would not exist a star. The only mode, therefore, in which, under such a state of affairs, we could comprehend the voids which our telescopes find in innumerable directions, would be by supposing the distance of the invisible background so immense that no ray from it has yet been able to reach us at all."

2.2 ■ ON LARGE SCALES, THE UNIVERSE IS ISOTROPIC AND HOMOGENEOUS

What does it mean to state that the universe is isotropic and homogeneous? Saying that the universe is *isotropic* means that there are no preferred directions in the universe; it looks the same no matter which way you point your telescope. Saying that the universe is *homogeneous* means that there are no preferred locations in the universe; it looks the same no matter where you set up your telescope. Note the very important qualifier: the universe is isotropic and homogeneous *on large scales*. In this context, large scales means that the universe is only isotropic and homogeneous on scales of roughly 100 Mpc or more.

The isotropy of the universe is not immediately obvious. In fact, on small scales, the universe is blatantly anisotropic. Consider, for example, a sphere 3 m in diameter, centered on your navel (Figure 2.2a). Within this sphere, there is a preferred direction; it is the direction commonly referred to as "down." It is easy to determine the vector pointing down. Just let go of a small dense object. The object doesn't hover in midair, and it doesn't move in a random direction; it falls down, toward the center of the Earth.

On significantly larger scales, the universe is still anisotropic. Consider, for example, a sphere 3 AU in diameter, centered on your navel (Figure 2.2b). Within this sphere, there is a preferred direction; it is the direction pointing toward the Sun, which is by far the most massive and most luminous object within the sphere. It is easy to determine the vector pointing toward the Sun. Just step outside on a sunny day, and point to that really bright disk of light up in the sky.

On still larger scales, the universe is *still* anisotropic. Consider, for example, a sphere 3 Mpc in diameter, centered on your navel (Figure 2.2c). This sphere contains the Local Group of galaxies, a small cluster of some 40 galaxies. By far the most massive and most luminous galaxies in the Local Group are our own galaxy and M31, which together contribute about 86% of the total luminosity within the 3 Mpc sphere. Thus, within this sphere, our galaxy and M31 define a preferred direction. It is fairly easy to determine the vector pointing from our galaxy to M31; just step outside on a clear night when the constellation Andromeda is above the horizon, and point to the fuzzy oval in the middle of the constellation.

It isn't until you get to considerably larger scales that the universe can be considered as isotropic. Consider a sphere 200 Mpc in diameter, centered on your navel. Figure 2.2d shows a slice through such a sphere, with superclusters of

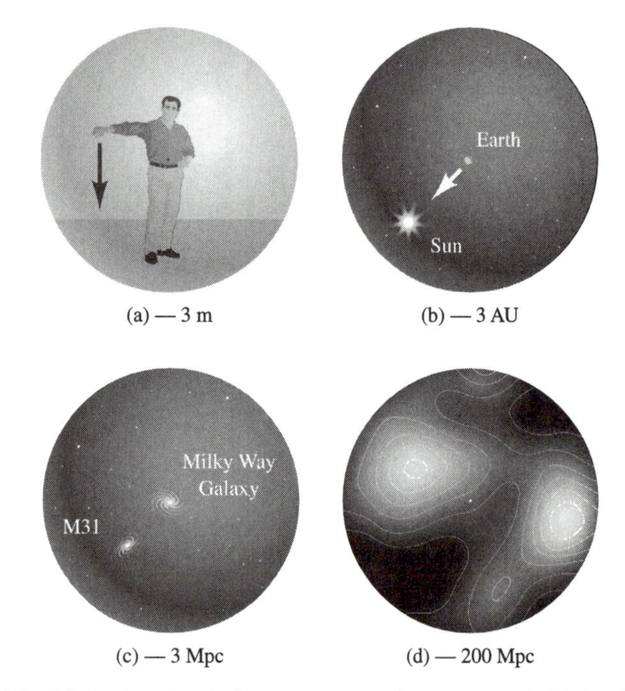

FIGURE 2.2 (a) A sphere 3 m in diameter, centered on your navel. (b) A sphere 3 AU in diameter, centered on your navel. (c) A sphere 3 Mpc in diameter, centered on your navel. (d) A sphere 200 Mpc in diameter, centered on your navel. Shown is the number density of galaxies smoothed with a Gaussian of width 17 Mpc. The heavy contour is drawn at the mean density; darker regions represent higher density, lighter regions represent lower density.

galaxies indicated as light patches. The Perseus–Pisces supercluster is on the right, the Hydra–Centaurus supercluster is on the left, and the edge of the Coma supercluster is just visible at the top of Figure 2.2d. Superclusters are typically $\sim 100\,\mathrm{Mpc}$ along their longest dimensions, and are separated by voids (low density regions) which are typically $\sim 100\,\mathrm{Mpc}$ across. These are the largest structures in the universe, it seems; surveys of the universe on still larger scales don't find "superduperclusters."

On small scales, the universe is obviously inhomogeneous, or lumpy, in addition to being anisotropic. For instance, a sphere 3 m in diameter, centered on your navel, will have an average density of $\sim 100\,\mathrm{kg\,m^{-3}}$, in round numbers. However, the average density of the universe as a whole is $\rho_0 \sim 3 \times 10^{-27}\,\mathrm{kg\,m^{-3}}$. Thus, on a scale $d \sim 3\,\mathrm{m}$, the patch of the universe surrounding you is more than 28 orders of magnitude denser than average.

On significantly larger scales, the universe is still inhomogeneous. A sphere 3 AU in diameter, centered on your navel, has an average density of $4 \times 10^{-5}\,\mathrm{kg\,m^{-3}}$; that's 22 orders of magnitude denser than the average for the universe.

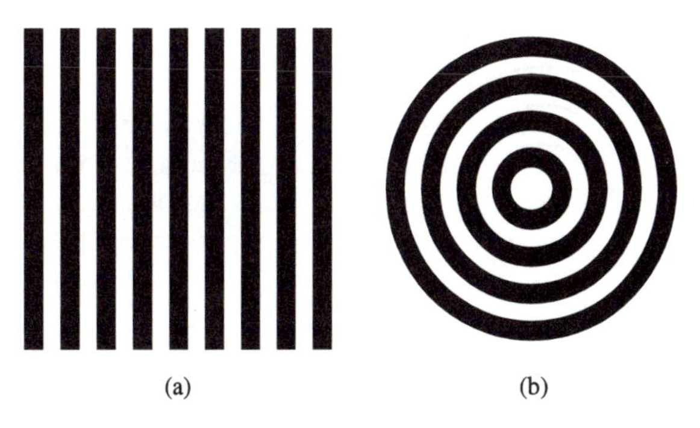

FIGURE 2.3 (a) A pattern that is anisotropic, but is homogeneous on scales larger than the stripe width. (b) A pattern that is isotropic about the origin, but is inhomogeneous.

On still larger scales, the universe is *still* inhomogeneous. A sphere 3 Mpc in diameter, centered on your navel, will have an average density of $\sim 3 \times 10^{-26}\,\text{kg}\,\text{m}^{-3}$, still an order of magnitude denser than the universe as a whole. It's only when you contemplate a sphere $\sim$ 100 Mpc in diameter that a sphere centered on your navel is not overdense compared to the universe as a whole.

Note that homogeneity does not imply isotropy. A sheet of paper printed with stripes (Figure 2.3a) is homogeneous on scales larger than the stripe width, but it is not isotropic. The direction of the stripes provides a preferred direction by which you can orient yourself. Note also that isotropy around a single point does not imply homogeneity. A sheet of paper printed with a bullseye (Figure 2.3b) is isotropic around the center of the bullseye, but is it not homogeneous. The rings of the bullseye look different far from the center than they do close to the center. You can tell where you are relative to the center by measuring the radius of curvature of the nearest ring.

In general, then, saying that something is homogeneous is quite different from saying it is isotropic. However, modern cosmologists have adopted the *cosmological principle*, which states "There is nothing special about our location in the universe." The cosmological principle holds true only on large scales (of 100 Mpc or more). On smaller scales, your navel obviously is in a special location. Most spheres 3 m across don't contain a sentient being; most spheres 3 AU across don't contain a star; most spheres 3 Mpc across don't contain a pair of bright galaxies. However, most spheres over 100 Mpc across do contain roughly the same pattern of superclusters and voids, statistically speaking. The universe, on scales of 100 Mpc or more, appears to be isotropic around us. Isotropy around any point in the universe, such as your navel, combined with the cosmological principle, implies isotropy around every point in the universe; and isotropy around every point in the universe *does* imply homogeneity.

The cosmological principle has the alternate name of the "Copernican principle" as a tribute to Copernicus, who pointed out that the Earth is not the center of the universe. Later cosmologists also pointed out the Sun is not the center, that our galaxy is not the center, and that the Local Group is not the center. In fact, there is no center to the universe.

2.3 ■ GALAXIES SHOW A REDSHIFT PROPORTIONAL TO THEIR DISTANCE

When we look at a galaxy at visible wavelengths, we detect primarily the light from the stars that the galaxy contains. Thus, when we take a galaxy's spectrum at visible wavelengths, it typically contains absorption lines created in the stars' relatively cool upper atmospheres.[3] Suppose we consider a particular absorption line whose wavelength, as measured in a laboratory here on Earth, is λ_{em}. The wavelength we measure for the same absorption line in a distant galaxy's spectrum, λ_{ob}, will not, in general, be the same. We say that the galaxy has a redshift z, given by the formula

$$z \equiv \frac{\lambda_{ob} - \lambda_{em}}{\lambda_{em}}. \tag{2.4}$$

Strictly speaking, when $z < 0$, this quantity is called a blueshift, rather than a redshift. However, the vast majority of galaxies have $z > 0$.

The fact that the light from galaxies is generally redshifted to longer wavelengths, rather than blueshifted to shorter wavelengths, was not known until the twentieth century. In 1912, Vesto Slipher at the Lowell Observatory measured the shift in wavelength of the light from M31; this galaxy, as it turns out, is one of the few that exhibits a blueshift. By 1925, Slipher had measured the shifts in the spectral lines for approximately 40 galaxies, finding that they were nearly all redshifted; the exceptions were all nearby galaxies within the Local Group.

By 1929, enough galaxy redshifts had been measured for the cosmologist Edwin Hubble to make a study of whether a galaxy's redshift depends on its distance from us. Although measuring a galaxy's redshift is relatively easy, and can be done with high precision, measuring its distance is difficult. Hubble knew z for nearly 50 galaxies, but had estimated distances for only 20 of them. Nevertheless, from a plot of redshift (z) versus distance (r), reproduced in Figure 2.4, he found the famous linear relation now known as Hubble's Law:

$$z = \frac{H_0}{c} r, \tag{2.5}$$

where H_0 is a constant (now called the Hubble constant). Hubble interpreted the observed redshift of galaxies as being a Doppler shift due to their radial velocity away from Earth. Because the values of z in Hubble's analysis were all small ($z < 0.004$), he was able to use the classical, nonrelativistic relation for the Doppler shift, $z = v/c$, where v is the radial velocity of the light source (in this case, a galaxy). Interpreting the redshifts as Doppler shifts, Hubble's law takes the form

$$v = H_0 r. \tag{2.6}$$

The Hubble constant H_0 can be found by dividing velocity by distance, so it is customarily written in the rather baroque units of $\text{km}\,\text{s}^{-1}\,\text{Mpc}^{-1}$. When Hubble

[3] Galaxies containing active galactic nuclei will also show *emission* lines from the hot gas in their nuclei.

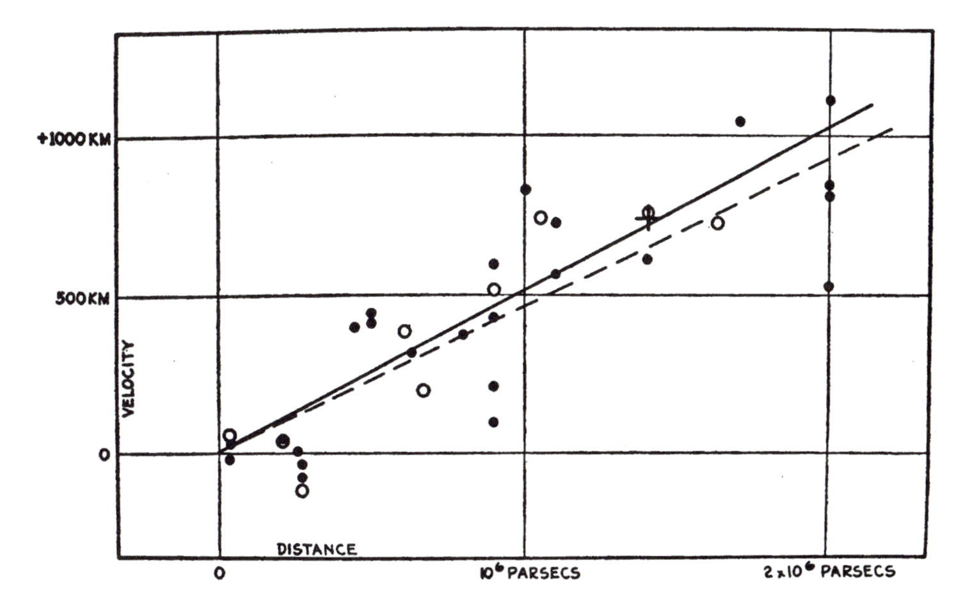

FIGURE 2.4 Edwin Hubble's original plot of the relation between redshift (vertical axis) and distance (horizontal axis). Note that the vertical axis actually plots cz rather than z, and that the units are accidentally written as km rather than km/s.

first discovered Hubble's Law, he thought that the numerical value of the Hubble constant was $H_0 = 500\,\mathrm{km\,s^{-1}\,Mpc^{-1}}$ (see Figure 2.4). However, it turned out that Hubble was severely underestimating the distances to galaxies.

Figure 2.5 shows a more recent determination of the Hubble constant from nearby galaxies, using data obtained by (appropriately enough) the Hubble Space Telescope. The best current estimate of the Hubble constant, combining the results of different research groups, is

$$H_0 = 70 \pm 7\,\mathrm{km\,s^{-1}\,Mpc^{-1}}. \tag{2.7}$$

This is the value for the Hubble constant that we will use in the remainder of this book.

Cosmological innocents sometimes exclaim, when first encountering Hubble's Law, "Surely it must be a violation of the cosmological principle to have all those distant galaxies moving away from *us*! It looks as if we are at a special location in the universe—the point away from which all other galaxies are fleeing." In fact, what we see here in our galaxy is exactly what you would expect to see in a universe that is undergoing homogeneous and isotropic expansion. We see distant galaxies moving away from us; but observers in any other galaxy would also see distant galaxies moving away from them.

To see on a more mathematical level what we mean by homogeneous, isotropic expansion, consider three galaxies at positions $\vec{r}_1$, $\vec{r}_2$, and $\vec{r}_3$. They define a triangle

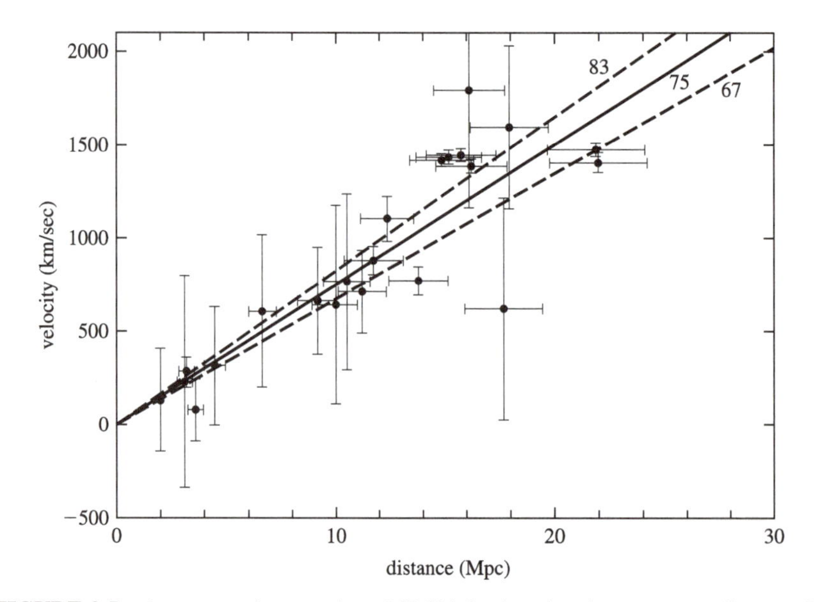

FIGURE 2.5 A more modern version of Hubble's plot, showing cz versus distance. In this case, the galaxy distances have been determined using Cepheid variable stars as standard candles, as described in Chapter 6.

(Figure 2.6) with sides of length

$$r_{12} \equiv |\vec{r}_1 - \vec{r}_2| \tag{2.8}$$

$$r_{23} \equiv |\vec{r}_2 - \vec{r}_3| \tag{2.9}$$

$$r_{31} \equiv |\vec{r}_3 - \vec{r}_1| \, . \tag{2.10}$$

Homogeneous and uniform expansion means that the shape of the triangle is preserved as the galaxies move away from each other. Maintaining the correct relative lengths for the sides of the triangle requires an expansion law of the form

$$r_{12}(t) = a(t) r_{12}(t_0) \tag{2.11}$$

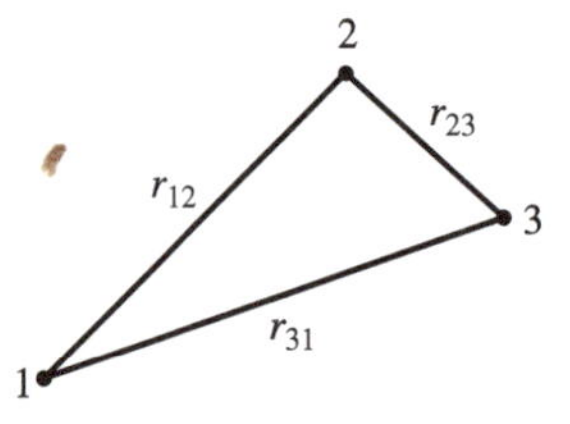

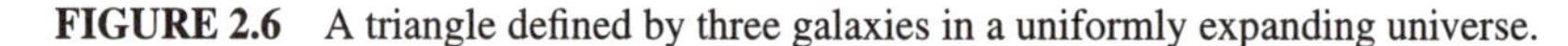

FIGURE 2.6 A triangle defined by three galaxies in a uniformly expanding universe.

$$r_{23}(t) = a(t)r_{23}(t_0) \tag{2.12}$$

$$r_{31}(t) = a(t)r_{31}(t_0). \tag{2.13}$$

Here the function $a(t)$ is a *scale factor*, equal to one at the present moment ($t = t_0$) and totally independent of location or direction. The scale factor $a(t)$ tells us how the expansion (or possibly contraction) of the universe depends on time. At any time t, an observer in galaxy 1 will see the other galaxies receding with a speed

$$v_{12}(t) = \frac{dr_{12}}{dt} = \dot{a}r_{12}(t_0) = \frac{\dot{a}}{a}r_{12}(t) \tag{2.14}$$

$$v_{31}(t) = \frac{dr_{31}}{dt} = \dot{a}r_{31}(t_0) = \frac{\dot{a}}{a}r_{31}(t). \tag{2.15}$$

You can demonstrate easily that an observer in galaxy 2 or galaxy 3 will find the same linear relation between observed recession speed and distance, with $\dot{a}/a$ playing the role of the Hubble constant. Since this argument can be applied to any trio of galaxies, it implies that in any universe where the distribution of galaxies is undergoing homogeneous, isotropic expansion, the velocity–distance relation takes the linear form $v = Hr$, with $H = \dot{a}/a$.

If galaxies are currently moving away from each other, then it implies they were closer together in the past. Consider a pair of galaxies currently separated by a distance r, with a velocity $v = H_0 r$ relative to each other. If there are no forces acting to accelerate or decelerate their relative motion, then their velocity is constant, and the time that has elapsed since they were in contact is

$$t_0 = \frac{r}{v} = \frac{r}{H_0 r} = H_0^{-1}, \tag{2.16}$$

independent of the current separation r. The time H_0^{-1} is referred to as the *Hubble time*. For $H_0 = 70 \pm 7\,\mathrm{km\,s^{-1}\,Mpc^{-1}}$, the Hubble time is $H_0^{-1} = 14.0 \pm 1.4\,\mathrm{Gyr}$. If the relative velocities of galaxies have been constant in the past, then one Hubble time ago, all the galaxies in the universe were crammed together into a small volume. Thus, the observation of galactic redshifts lead naturally to a *Big Bang* model for the evolution of the universe. A Big Bang model may be broadly defined as a model in which the universe expands from an initially highly dense state to its current low-density state.

The Hubble time of $\sim$ 14 Gyr is comparable to the ages computed for the oldest known stars in the universe. This rough equivalence is reassuring. However, the age of the universe—that is, the time elapsed since its original highly dense state—is not necessarily exactly equal to the Hubble time. We know that gravity exists, and that galaxies contain matter. If gravity working on matter is the only force at work on large scales, then the attractive force of gravity will act to slow the expansion. In this case, the universe was expanding more rapidly in the past than it is now, and the universe is younger than H_0^{-1}. On the other hand, if the energy density of the universe is dominated by a cosmological constant (an entity we'll examine in more detail in Chapter 4), then the dominant gravitational force is repulsive, and the universe may be older than H_0^{-1}.

Just as Hubble time provides a natural time scale for our universe, the Hubble distance, $c/H_0 = 4300 \pm 400\,\text{Mpc}$, provides a natural distance scale. Just as the age of the universe is roughly equal to H_0^{-1} in most Big Bang models, with the exact value depending on the expansion history of the universe, so the horizon distance (the greatest distance a photon can travel during the age of the universe) is roughly equal to c/H_0, with the exact value, again, depending on the expansion history. (Later chapters will deal with computing the exact values of the age and horizon size of our universe.)

Note how Hubble's Law ties in with Olbers' Paradox. If the universe is of finite age, $t_0 \sim H_0^{-1}$, then the night sky can be dark, even if the universe is infinitely large, because light from distant galaxies has not yet had time to reach us. Galaxy surveys tell us that the luminosity density of galaxies in the local universe is

$$nL \approx 2 \times 10^8\,\text{L}_\odot\,\text{Mpc}^{-3}. \tag{2.17}$$

By terrestrial standards, the universe is not a well-lit place; this luminosity density is equivalent to a single 40 watt light bulb within a sphere 1 AU in radius. If the horizon distance is $d_{\text{hor}} \sim c/H_0$, then the total flux of light we receive from all the stars from all the galaxies within the horizon will be

$$\begin{aligned} F_{\text{gal}} = 4\pi J_{\text{gal}} \approx nL \int_0^{r_H} dr \sim nL\left(\frac{c}{H_0}\right) \\ \sim 9 \times 10^{11}\,\text{L}_\odot\,\text{Mpc}^{-2} \sim 2 \times 10^{-11}\,\text{L}_\odot\,\text{AU}^{-2}. \end{aligned} \tag{2.18}$$

By the cosmological principle, this is the total flux of starlight you would expect at any randomly located spot in the universe. Comparing this to the flux we receive from the Sun,

$$F_{\text{sun}} = \frac{1\,\text{L}_\odot}{4\pi\,\text{AU}^2} \approx 0.08\,\text{L}_\odot\,\text{AU}^{-2}, \tag{2.19}$$

we find that $F_{\text{gal}}/F_{\text{sun}} \sim 3 \times 10^{-10}$. Thus, the total flux of starlight at a randomly selected location in the universe is less than a billionth the flux of light we receive from the Sun here on Earth. For the entire universe to be as well-lit as the Earth, it would have to be over a billion times older than it is; *and* you'd have to keep the stars shining during all that time.

Hubble's Law occurs naturally in a Big Bang model for the universe, in which homogeneous and isotropic expansion causes the density of the universe to decrease steadily from its initial high value. In a Big Bang model, the properties of the universe evolve with time; the average density decreases, the mean distance between galaxies increases, and so forth. However, Hubble's Law can also be explained by a *Steady State* model. The Steady State model was first proposed in the 1940's by Hermann Bondi, Thomas Gold, and Fred Hoyle, who were proponents of the *perfect cosmological principle*, which states that not only are there no privileged locations in space, there are no privileged moments in time. Thus, a Steady State universe is one in which the global properties of the universe, such as the mean density ρ_0 and the Hubble constant H_0, remain constant with time.

In a Steady State universe, the velocity–distance relation

$$\frac{dr}{dt} = H_0 r \tag{2.20}$$

can be easily integrated, since H_0 is constant with time, to yield an exponential law:

$$r(t) \propto e^{H_0 t}. \tag{2.21}$$

Note that $r \to 0$ only in the limit $t \to -\infty$; a Steady State universe is infinitely old. If there existed an instant in time at which the universe started expanding (as in a Big Bang model), that would be a special moment, in violation of the assumed "perfect cosmological principle." The volume of a spherical region of space, in a Steady State model, increases exponentially with time:

$$V = \frac{4\pi}{3} r^3 \propto e^{3H_0 t}. \tag{2.22}$$

However, if the universe is in a steady state, the density of the sphere must remain constant. To have a constant density of matter within a growing volume, matter must be continuously created at a rate

$$\dot{M}_{\rm ss} = \rho_0 \dot{V} = \rho_0 3 H_0 V. \tag{2.23}$$

If our own universe, with matter density $\rho_0 \sim 3 \times 10^{-27}\,\mathrm{kg\,m^{-3}}$, happened to be a Steady State universe, then matter would have to be created at a rate

$$\frac{\dot{M}_{\rm ss}}{V} = 3 H_0 \rho_0 \sim 6 \times 10^{-28}\,\mathrm{kg\,m^{-3}\,Gyr^{-1}}. \tag{2.24}$$

This corresponds to creating roughly one hydrogen atom per cubic kilometer per year.

During the 1950s and 1960s, the Big Bang and Steady State models battled for supremacy. Critics of the Steady State model pointed out that the continuous creation of matter violates mass-energy conservation. Supporters of the Steady State model pointed out that the continuous creation of matter is no more absurd than the instantaneous creation of the entire universe in a single "Big Bang" billions of years ago.[4] The Steady State model finally fell out of favor when observational evidence increasingly indicated that the perfect cosmological principle is not true. The properties of the universe *do*, in fact, change with time. The discovery of the

[4]The name "Big Bang" was actually coined by Fred Hoyle, a supporter of the Steady State model.

Cosmic Microwave Background, discussed in section 2.5, is commonly regarded as the observation that decisively tipped the scales in favor of the Big Bang model.

2.4 ■ THE UNIVERSE CONTAINS DIFFERENT TYPES OF PARTICLES

It doesn't take a brilliant observer to confirm that the universe contains a variety of different things: shoes, ships, sealing wax, cabbages, kings, galaxies, and what have you. From a cosmologist's viewpoint, though, cabbages and kings are nearly indistinguishable—the main difference between them is that the mean mass per king is greater than the mean mass per cabbage. From a cosmological viewpoint, the most significant difference between the different components of the universe is that they are made of different elementary particles. The properties of the most cosmologically important particles are summarized in Table 2.1.

The material objects that surround us in our everyday life are made up of *protons*, *neutrons*, and *electrons*.[5] Protons and neutrons are examples of *baryons*, where a baryon is defined as a particle made of three quarks. A proton (p) contains two "up" quarks, each with an electrical charge of $+\frac{2}{3}$, and a "down" quark, with charge $-\frac{1}{3}$. A neutron (n) contains one "up" quark and two "down" quarks. Thus a proton has a net positive charge of $+1$, while a neutron is electrically neutral. Protons and neutrons also differ in their mass—or equivalently, in their rest energies. The proton mass is $m_p c^2 = 938.3\,\text{MeV}$, while the neutron mass is $m_n c^2 = 939.6\,\text{MeV}$, about 0.1% greater. Free neutrons are unstable, decaying into protons with a decay time of $\tau_n = 890\,\text{s}$, about a quarter of an hour. By contrast, experiments have put a lower limit on the decay time of the proton, which is very much greater than the Hubble time. Neutrons can be preserved against decay by binding them into an atomic nucleus with one or more protons.

Electrons (e^-) are examples of *leptons*, a class of elementary particles that are not made of quarks. The mass of an electron is much smaller than that of a neutron or proton; the rest energy of an electron is $m_e c^2 = 0.511\,\text{MeV}$. An electron has an electric charge equal in magnitude to that of a proton, but opposite in sign. On large scales, the universe is electrically neutral; the number of electrons is equal to the number of protons. Since protons outmass electrons by a factor of

TABLE 2.1 Particle Properties

Particle	Symbol	Rest energy (MeV)	Charge
proton	p	938.3	$+1$
neutron	n	939.6	0
electron	e^-	0.511	-1
neutrino	ν_e, ν_μ, ν_τ	?	0
photon	γ	0	0
dark matter	?	?	0

[5]For that matter, we ourselves are made of protons, neutrons, and electrons.

1836 to 1, the mass density of electrons is only a small perturbation to the mass density of protons and neutrons. For this reason, the component of the universe made up of ions, atoms, and molecules is generally referred to as *baryonic matter*, since only the baryons (protons and neutrons) contribute significantly to the mass density. Protons and neutrons are 800-pound gorillas; electrons are only 7-ounce bushbabies.

About three-fourths of the baryonic matter in the universe is currently in the form of ordinary hydrogen, the simplest of all elements. In addition, when we look at the remainder of the baryonic matter, it is primarily in the form of helium, the next simplest element. The Sun's atmosphere, for instance, contains 70% hydrogen by mass, and 28% helium; only 2% is contributed by more massive atoms. When astronomers look at a wide range of astronomical objects—stars and interstellar gas clouds, for instance—they find a minimum helium mass fraction of 24%. The baryonic component of the universe can be described, to lowest order, as a mix of three parts hydrogen to one part helium, with only minor contamination by heavier elements.

Another type of lepton, in addition to the electron, is the *neutrino* (ν). The most poetic summary of the properties of the neutrino was made by John Updike, in his poem "Cosmic Gall":[6]

> Neutrinos, they are very small.
> They have no charge and have no mass
> And do not interact at all.
> The earth is just a silly ball
> To them, through which they simply pass,
> Like dustmaids down a drafty hall
> Or photons through a sheet of glass.

In truth, Updike was using a bit of poetic license here. It is definitely true that neutrinos have no charge.[7] However, it is not true that neutrinos "do not interact at all"; they actually are able to interact with other particles via the weak nuclear force. The weak nuclear force, though, is very weak indeed; a typical neutrino emitted by the Sun would have to pass through a few parsecs of solid lead before having a 50% chance of interacting with a lead atom. Since neutrinos pass through neutrino detectors with the same facility with which they pass through the Earth, detecting neutrinos from astronomical sources is difficult.

There are three types, or "flavors," of neutrinos: electron neutrinos, muon neutrinos, and tau neutrinos. What Updike didn't know in 1960, when he wrote his poem, is that some or all of the neutrino types probably have a small mass. The evidence for massive neutrinos comes indirectly, from the search for neutrino oscillations. An *oscillation* is the transmutation of one flavor of neutrino into another. The rate at which two neutrino flavors oscillate is proportional to the difference

[6]From COLLECTED POEMS 1953–1993 by John Updike, ©1993 by John Updike. Used by permission of Alfred A. Knopf, a division of Random House, Inc.

[7]Their name, given them by Enrico Fermi, means "little neutral one" in Italian.

of the squares of their masses. Observations of neutrinos from the Sun are most easily explained if electron neutrinos (the flavor emitted by the Sun) oscillate into some other flavor of neutrino, with the difference in the squares of their masses being $\Delta(m_\nu^2 c^4) \approx 5 \times 10^{-5}\,\text{eV}^2$. Observations of muon neutrinos created by cosmic rays striking the upper atmosphere indicate that muon neutrinos oscillate into tau neutrinos, with $\Delta(m_\nu^2 c^4) \approx 3 \times 10^{-3}\,\text{eV}^2$ for these two flavors. Unfortunately, knowing the differences of the squares of the masses doesn't tell us the values of the masses themselves.

A particle which is known to be massless is the *photon*. Electromagnetic radiation can be thought of either as a wave or as a stream of particles, called photons. Light, when regarded as a wave, is characterized by its frequency f or its wavelength $\lambda = c/f$. When light is regarded as a stream of photons, each photon is characterized by its energy, $E_\gamma = hf$, where $h = 2\pi\hbar$ is the Planck constant. Photons of a wide range of energy, from radio to gamma rays, pervade the universe. Unlike neutrinos, photons interact readily with electrons, protons, and neutrons. For instance, photons can ionize an atom by kicking an electron out of its orbit, a process known as *photoionization*. Higher-energy photons can break an atomic nucleus apart, a process known as *photodissociation*.

Photons, in general, are easily created. One way to make photons is to take a dense, opaque object—such as the filament of an incandescent lightbulb—and heat it up. If an object is opaque, then the protons, neutrons, electrons, and photons that it contains frequently interact, and attain thermal equilibrium. When a system is in thermal equilibrium, the density of photons in the system, as a function of photon energy, depends only on the temperature T. It doesn't matter whether the system is a tungsten filament, an ingot of steel, or a sphere of ionized hydrogen and helium. The energy density of photons in the frequency range $f \rightarrow f + df$ is given by the *blackbody* function

$$\varepsilon(f)df = \frac{8\pi h}{c^3} \frac{f^3\,df}{\exp(hf/kT) - 1}, \tag{2.25}$$

illustrated in Figure 2.7. The peak in the blackbody function occurs at $hf_{\text{peak}} \approx 2.82kT$. Integrated over all frequencies, equation (2.25) yields a total energy density for blackbody radiation of

$$\varepsilon_\gamma = \alpha T^4, \tag{2.26}$$

where

$$\alpha = \frac{\pi^2}{15} \frac{k^4}{\hbar^3 c^3} = 7.56 \times 10^{-16}\,\text{J}\,\text{m}^{-3}\,\text{K}^{-4}. \tag{2.27}$$

The number density of photons in blackbody radiation can be computed from equation (2.25) as

$$n_\gamma = \beta T^3, \tag{2.28}$$

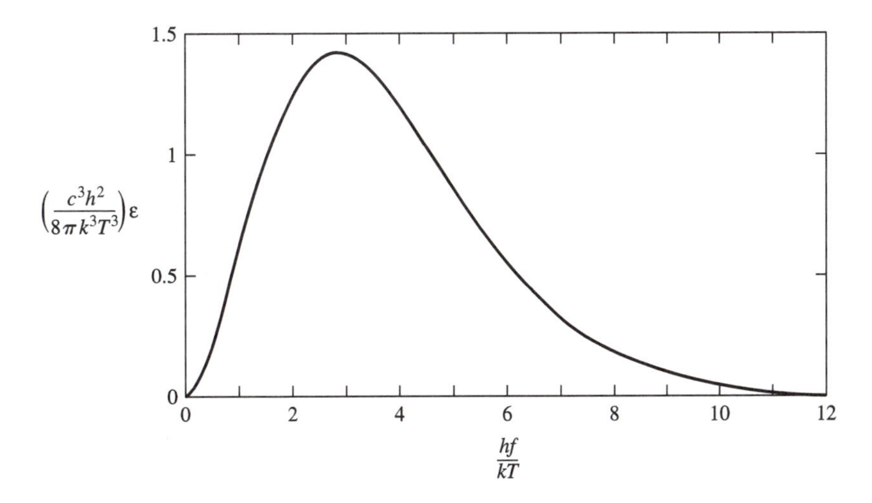

FIGURE 2.7 The energy distribution of a blackbody spectrum.

where

$$\beta = \frac{2.404}{\pi^2}\frac{k^3}{\hbar^3 c^3} = 2.03 \times 10^7\,\text{m}^{-3}\,\text{K}^{-3}. \tag{2.29}$$

Division of equation (2.26) by equation (2.28) yields a mean photon energy of $E_{\text{mean}} = hf_{\text{mean}} \approx 2.70kT$, close to the peak in the spectrum. You have a temperature of 310 K, and you radiate an approximate blackbody spectrum, with a mean photon energy of $E_{\text{mean}} \approx 0.072\,\text{eV}$, corresponding to a wavelength of $\lambda \approx 1.7 \times 10^{-5}\,\text{m}$, in the infrared. By contrast, the Sun produces an approximate blackbody spectrum with a temperature $T_\odot \approx 5800\,\text{K}$. This implies a mean photon energy $E_{\text{mean}} \approx 1.3\,\text{eV}$, corresponding to $\lambda \approx 9.0 \times 10^{-7}\,\text{m}$, in the near infrared. Note, however, that although the mean photon energy in a blackbody spectrum is $\sim 3kT$, Figure 2.7 shows us that there is a long exponential tail to higher photon energies. A large fraction of the Sun's output is at wavelengths of $(4 \rightarrow 7) \times 10^{-7}\,\text{m}$, which our eyes are equipped to detect.

The most mysterious component of the universe is *dark matter*. When observational astronomers refer to dark matter, they usually mean any massive component of the universe that is too dim to be detected readily using current technology. Thus, stellar remnants such as white dwarfs, neutron stars, and black holes are sometimes referred to as dark matter, since an isolated stellar remnant is extremely faint and difficult to detect. Substellar objects such as brown dwarfs are also referred to as dark matter, since brown dwarfs, too low in mass for nuclear fusion to occur in their cores, are very dim. Theoretical astronomers sometimes use a more stringent definition of dark matter than do observers, defining dark matter as any massive component of the universe which doesn't emit, absorb, or scatter light at all.[8] If neutrinos have mass, for instance, as the recent neutrino oscillation results indicate, they qualify as dark matter. In some extensions to the Standard Model of particle physics, there exist massive particles that interact, like neutrinos, only through the weak nuclear force and through gravity. These particles, which have not yet been detected in the laboratory, are generically referred to as Weakly Interacting Massive Particles, or WIMPs.

In this book, we will generally adopt the broader definition of dark matter as something which is too dim for us to see, even with our best available technology. Detecting dark matter is, naturally, difficult. The standard method of detecting dark matter is by measuring its gravitational effect on luminous matter, just as the planet Neptune was first detected by its gravitational effect on the planet Uranus. Although Neptune no longer qualifies as dark matter, observations of the motions of stars within galaxies and of galaxies within clusters indicate that a significant amount of dark matter is in the universe. Exactly how much there is, and what it's made of, is a topic of great interest to cosmologists.

2.5 ■ THE UNIVERSE IS FILLED WITH A COSMIC MICROWAVE BACKGROUND

The discovery of the Cosmic Microwave Background (CMB) by Arno Penzias and Robert Wilson in 1965 has entered cosmological folklore. Using a microwave antenna at Bell Labs, they found an isotropic background of microwave radiation. More recently, the Cosmic Background Explorer (COBE) satellite has revealed that the Cosmic Microwave Background is exquisitely well fitted by a blackbody spectrum (equation (2.25)) with a temperature

$$T_0 = 2.725 \pm 0.001\,\mathrm{K}. \tag{2.30}$$

The energy density of the CMB is, from equation (2.26),

$$\varepsilon_\gamma = 4.17 \times 10^{-14}\,\mathrm{J\,m^{-3}}. \tag{2.31}$$

This is roughly equivalent to a quarter of an MeV per cubic meter of space. The number density of CMB photons is, from equation (2.28),

$$n_\gamma = 4.11 \times 10^{8}\,\mathrm{m^{-3}}. \tag{2.32}$$

Thus, there are about 411 CMB photons in every cubic centimeter of the universe at the present day. The mean energy of CMB photons, however, is quite low, only

$$E_{\mathrm{mean}} = 6.34 \times 10^{-4}\,\mathrm{eV}. \tag{2.33}$$

This is too low in energy to photoionize an atom, much less photodissociate a nucleus. About all they do, from a terrestrial point of view, is cause static on

[8] Using this definition, an alternate name for dark matter might be "transparent matter" or "invisible matter." However, the name "dark matter" has received the sanction of history.

television. The mean CMB photon energy corresponds to a wavelength of 2 millimeters, in the microwave region of the electromagnetic spectrum—hence the name "Cosmic *Microwave* Background."

The existence of the CMB is a very important cosmological clue. In particular, it is the clue that caused the Big Bang model for the universe to be favored over the Steady State model. In a Steady State universe, the existence of blackbody radiation at 2.725 K is not easily explained. In a Big Bang universe, however, a cosmic background radiation arises naturally if the universe was initially very hot as well as very dense. If mass is conserved in an expanding universe, then in the past the universe was denser than it is now. Assume that the early dense universe was very hot ($T \gg 10^4$ K, or $kT \gg 1\,\mathrm{e}V$). At such high temperatures, the baryonic matter in the universe was completely ionized, and the free electrons rendered the universe opaque. A dense, hot, opaque body, as described in Section 2.4, produces blackbody radiation. So, the early hot dense universe was full of photons, banging off the electrons like balls in a pinball machine, with a spectrum typical of a blackbody (equation (2.25)). However, as the universe expanded, it cooled. When the temperature dropped to ~ 3000 K, ions and electrons combined to form neutral atoms. When the universe no longer contained a significant number of free electrons, the blackbody photons started streaming freely through the universe, without further scattering off free electrons.

The blackbody radiation that fills the universe today can be explained as a relic of the time when the universe was sufficiently hot and dense to be opaque. However, at the time the universe became transparent, its temperature was ~ 3000 K. The temperature of the CMB today is 2.725 K, a factor of 1100 lower. The drop in temperature of the blackbody radiation is a direct consequence of the expansion of the universe. Consider a region of volume V that expands at the same rate as the universe, so that $V \propto a(t)^3$. The blackbody radiation in the volume can be thought of as a photon gas with energy density $\varepsilon_\gamma = \alpha T^4$. Moreover, since the photons in the volume have momentum as well as energy, the photon gas has a pressure; the pressure of a photon gas is $P_\gamma = \varepsilon_\gamma/3$. The photon gas within our imaginary box must follow the laws of thermodynamics; in particular, the boxful of photons must obey the first law

$$dQ = dE + PdV, \tag{2.34}$$

where dQ is the amount of heat flowing into or out of the photon gas in the volume V, dE is the change in the internal energy, P is the pressure, and dV is the change in volume of the box. Since, in a homogeneous universe, there is no net flow of heat (everything is the same temperature, after all), $dQ = 0$. Thus, the first law of thermodynamics, applied to an expanding homogeneous universe, is

$$\frac{dE}{dt} = -P(t)\frac{dV}{dt}. \tag{2.35}$$

Since, for the photons of the CMB, $E = \varepsilon_\gamma V = \alpha T^4 V$ and $P = P_\gamma = \alpha T^4/3$, equation (2.35) can be rewritten in the form

$$\alpha\left(4T^3\frac{dT}{dt}V + T^4\frac{dV}{dt}\right) = -\frac{1}{3}\alpha T^4\frac{dV}{dt}, \tag{2.36}$$

or

$$\frac{1}{T}\frac{dT}{dt} = -\frac{1}{3V}\frac{dV}{dt}. \tag{2.37}$$

However, since $V \propto a(t)^3$ as the box expands, this means that the rate in change of the photons' temperature is related to the rate of expansion of the universe by the relation

$$\frac{d}{dt}(\ln T) = -\frac{d}{dt}(\ln a). \tag{2.38}$$

This implies the simple relation $T(t) \propto a(t)^{-1}$; the temperature of the cosmic background radiation has dropped by a factor of 1100 since the universe became transparent, because the scale factor $a(t)$ has increased by a factor of 1100 since then. What we now see as a Cosmic Microwave Background was once, at the time the universe became transparent, a Cosmic *Near-Infrared* Background, with a temperature slightly cooler than the surface of the star Betelgeuse.

The evidence cited so far can all be explained within the framework of a *Hot Big Bang* model, in which the universe was originally very hot and very dense, and since then has been expanding and cooling. The remainder of this book will be devoted to working out the details of the Hot Big Bang model that best fits the universe in which we live.

SUGGESTED READING

Full references are given in the Annotated Bibliography on page 235.

Bernstein (1995): The "Micropedia" that begins this text is a useful overview of the contents of the universe and the forces which work on them

Harrison (1987): The definitive treatment of Olbers' paradox

PROBLEMS

2.1. Suppose that in Sherwood Forest, the average radius of a tree is $R = 1\,\text{m}$ and the average number of trees per unit area is $\Sigma = 0.005\,\text{m}^{-2}$. If Robin Hood shoots an arrow in a random direction, how far, on average, will it travel before it strikes a tree?

2.2. Suppose you are in an infinitely large, infinitely old universe in which the average density of stars is $n_\star = 10^9\,\text{Mpc}^{-3}$ and the average stellar radius is equal to the Sun's radius: $R_\star = R_\odot = 7 \times 10^8\,\text{m}$. How far, on average, could you see in any direction before your line of sight struck a star? (Assume standard Euclidean geometry holds true in this universe.) If the stars are clumped into galaxies with a density $n_g =$

$1\,\text{Mpc}^{-3}$ and average radius $R_g = 2000\,\text{pc}$, how far, on average, could you see in any direction before your line of sight hit a galaxy?

2.3. Since you are made mostly of water, you are very efficient at absorbing microwave photons. If you were in intergalactic space, approximately how many CMB photons would you absorb per second? (If you like, you may assume you are spherical.) What is the approximate rate, in watts, at which you would absorb radiative energy from the CMB? Ignoring other energy inputs and outputs, how long would it take the CMB to raise your temperature by one nanoKelvin (10^{-9} K)? (You may assume your heat capacity is the same as pure water, $C = 4200\,\text{J}\,\text{kg}^{-1}\,\text{K}^{-1}$.)

2.4. Suppose that the difference between the square of the mass of the electron neutrino and that of the muon neutrino has the value $[m(\nu_\mu)^2 - m(\nu_e)^2]c^4 = 5 \times 10^{-5}\,\text{eV}^2$, and that the difference between the square of the mass of the muon neutrino and that of the tau neutrino has the value $[m(\nu_\tau)^2 - m(\nu_\mu)^2]c^4 = 3 \times 10^{-3}\,\text{eV}^2$. (This is consistent with the observational results discussed in section 2.4.) What values of $m(\nu_e)$, $m(\nu_\mu)$, and $m(\nu_\tau)$ minimize the sum $m(\nu_e) + m(\nu_\mu) + m(\nu_\tau)$, given these constraints?

2.5. A hypothesis once used to explain the Hubble relation is the "tired light hypothesis." The tired light hypothesis states that the universe is not expanding, but that photons simply lose energy as they move through space (by some unexplained means), with the energy loss per unit distance being given by the law

$$\frac{dE}{dr} = -KE, \tag{2.39}$$

where K is a constant. Show that this hypothesis gives a distance-redshift relation that is linear in the limit $z \ll 1$. What must the value of K be in order to yield a Hubble constant of $H_0 = 70\,\text{km}\,\text{s}^{-1}\,\text{Mpc}^{-1}$?

CHAPTER 3

Newton Versus Einstein

On cosmological scales (that is, on scales greater than 100 Mpc or so), the dominant force determining the evolution of the universe is gravity. The weak and strong nuclear forces are short-range forces; the weak force is effective only on scales of $\ell_w \sim 10^{-18}$ m or less, and the strong force on scales of $\ell_s \sim 10^{-15}$ m or less. Both gravity and electromagnetism are long-range forces. On small scales, gravity is negligibly small compared to electromagnetic forces; for instance, the electrostatic repulsion between a pair of protons is larger by a factor $\sim 10^{36}$ than the gravitational attraction between them. However, on large scales, the universe is electrically neutral, so there are no electrostatic forces on large scales. Moreover, intergalactic magnetic fields are sufficiently small that magnetic forces are also negligibly tiny on cosmological scales. Ironically then, gravity—the weakest of all forces from a particle physics standpoint—is the force that determines the evolution of the universe on large scales.

Note that in referring to gravity as a force, we are implicitly adopting a Newtonian viewpoint. In physics, the two useful ways of looking at gravity are the Newtonian, or classical, viewpoint and the Einsteinian, or general relativistic, viewpoint. In Isaac Newton's view, as formulated by his Laws of Motion and Law of Gravity, gravity is a force that causes massive bodies to be accelerated. By contrast, in Einstein's view, gravity is a manifestation of the curvature of space-time. Although Newton's view and Einstein's view are conceptually very different, in most cosmological contexts they yield the same predictions. The Newtonian predictions differ significantly from the predictions of general relativity only in the limit of deep potential minima (to use Newtonian language) or strong spatial curvature (to use general relativistic language). In these limits, general relativity yields the correct result.

In the limit of shallow potential minima and weak spatial curvature, it is permissible to switch back and forth between a Newtonian and a general relativistic viewpoint, adopting whichever one is more convenient. I will frequently adopt the Newtonian view of gravity in this book because, in many contexts, it is mathematically simpler and conceptually more familiar. The question of *why* it is possible to switch back and forth between the two very different viewpoints of Newton and Einstein is an intriguing one, and deserves closer investigation.

3.1 ■ EQUIVALENCE PRINCIPLE

In Newton's view of the universe, space is unchanging and Euclidean. In Euclidean, or "flat," space, all the axioms and theorems of plane geometry (as codified by Euclid in the third century BC) hold true. In Euclidean space, the shortest distance between two points is a straight line, the angles at the vertices of a triangle sum to π radians, the circumference of a circle is 2π times its radius, and so on, through all the other axioms and theorems you learned in high school geometry. In Newton's view, moreover, an object with no net force acting on it moves in a straight line at constant speed. However, when we look at objects in the Solar System such as planets, moons, comets, and asteroids, we find that they move on curved lines, with constantly changing speed. Why is this? Newton would tell us, "Their velocities are changing because there is a force acting on them; the force called *gravity*."

Newton devised a formula for computing the gravitational force between two objects. Every object in the universe, said Newton, has a property that we may call the "gravitational mass." Let the gravitational masses of two objects be M_g and m_g, and let the distance between their centers be r. The gravitational force acting between the two objects (assuming they are both spherical) is

$$F = -\frac{GM_g m_g}{r^2}. \tag{3.1}$$

The negative sign in the above equation indicates that gravity, in the Newtonian view, is always an attractive force, tending to draw two bodies closer together.

What is the acceleration that results from this gravitational force? Newton had something to say about that as well. Every object in the universe, said Newton, has a property that we may call the "inertial mass." Let the inertial mass of an object be m_i. Newton's second law of motion says that force and acceleration are related by the equation

$$F = m_i a. \tag{3.2}$$

In equations (3.1) and (3.2) we have distinguished, through the use of different subscripts, between the gravitational mass m_g and the inertial mass m_i. One of the fundamental principles of physics is that the gravitational mass and the inertial mass of an object are identical:

$$m_g = m_i. \tag{3.3}$$

When you stop to think about it, this equality is a remarkable fact. The property of an object that determines how strongly it is pulled on by the force of gravity is equal to the property that determines its resistance to acceleration by *any* force, not just the force of gravity. The equality of gravitational mass and inertial mass is called the *equivalence principle*, and it is the equivalence principle that led Einstein to devise his theory of general relativity.

If the equivalence principle did not hold, then the gravitational acceleration of an object toward a mass M_g would be (combining equations (3.1) and (3.2))

$$a = -\frac{GM_g}{r^2}\left(\frac{m_g}{m_i}\right), \tag{3.4}$$

with the ratio m_g/m_i varying from object to object. However, when Galileo dropped objects from towers and slid objects down inclined planes, he found that the acceleration (barring the effects of air resistance and friction) was always the same, regardless of the mass and composition of the object. The magnitude of the gravitational acceleration close to the Earth's surface is $g = GM_{\rm Earth}/r_{\rm Earth}^2 = 9.8\,{\rm m\,s^{-2}}$. Modern tests of the equivalence principle, which are basically more sensitive versions of Galileo's experiments, reveal that the inertial and gravitational masses are the same to within one part in 10^{12}.

To see how the equivalence principle led Einstein to devise his theory of general relativity, let's begin with a thought experiment of the sort Einstein would devise. Suppose you wake up one morning to find that you have been sealed up (bed and all) within an opaque, soundproof, hermetically sealed box. "Oh no!" you say. "This is what I've always feared would happen. I've been abducted by space aliens who are taking me away to their home planet." So startled are you by this realization, you drop your teddy bear. Observing the falling bear, you find that it falls toward the floor of the box with an acceleration $a = 9.8\,{\rm m\,s^{-2}}$. "Whew!" you say, with some relief. "At least I am still on the Earth's surface; they haven't taken me away in their spaceship yet." At that moment, a window in the side of the box opens to reveal (much to your horror) that you are inside an alien spaceship that is being accelerated at $9.8\,{\rm m\,s^{-2}}$ by a rocket engine. When you drop a teddy bear, or any other object, within a sealed box, the equivalence principle permits two possible interpretations, with no way of distinguishing between them: (1) The box is static, or moving with a constant velocity, and the bear is being accelerated downward by a constant gravitational force; (2) The bear is static, or moving at a constant velocity, and the box is being accelerated upward at a constant rate. The behavior of the bear in each case (Figure 3.1) is identical. In each case, a big bear falls at the same rate as a little bear; in each case, a bear stuffed with cotton falls at the same rate as a bear stuffed with lead; and in each case, a sentient anglophone bear would say, "Oh, bother. I'm weightless," during the interval before it collides with the floor of the box.[1]

Einstein's insight, starting from the equivalence principle, led him to the theory of general relativity. To understand Einstein's thought processes, imagine yourself back in the sealed box, being accelerated through interplanetary space at

[1]Note that the equivalence of the two boxes in Figure 3.1 depends on the gravitational acceleration in the left-hand box being constant. In the real universe, though, gravitational accelerations are not exactly constant, but vary with position. For instance, the gravitational acceleration near the Earth's surface is a vector $\vec{g}(\vec{r})$ that varies in direction (always pointing toward the Earth's center) and in magnitude (decreasing as the inverse square of the distance from the Earth's center). Thus, in the real universe, the equivalence principle can only be applied to an infinitesimally small box—that is, a box so small that the variation in $\vec{g}$ is too tiny to be measured.

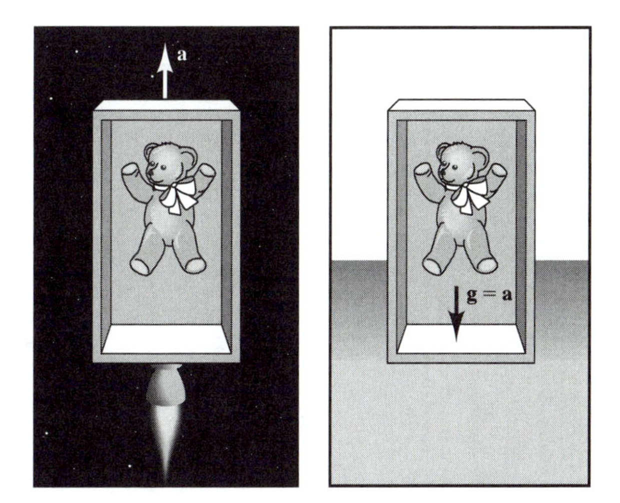

FIGURE 3.1 Equivalence principle (teddy bear version). The behavior of a bear in an accelerated box (left) is identical to that of a bear being accelerated by gravity (right).

$9.8\,\mathrm{m\,s^{-2}}$. You grab the flashlight that you keep on the bedside table and shine a beam of light perpendicular to the acceleration vector (Figure 3.2). Since the box is accelerating upward, the path of the light beam will appear to you to be bent downward, as the floor of the box rushes up to meet the photons. However, thanks to the equivalence principle, we can replace the accelerated box with a stationary box experiencing a constant gravitational acceleration. Since there's no way to distinguish between these two cases, we are led to the conclusion that the paths of photons will be curved downward in the presence of a gravitational field. Gravity affects photons, Einstein concluded, even though they have no mass. Contemplating the curved path of the light beam, Einstein had one more insight. One of the fundamental principles of optics is *Fermat's Principle*, which states that light travels between two points along a path which minimizes the travel time required.[2] In a vacuum, where the speed of light is constant, this translates into the requirement that light takes the shortest path between two points. In Euclidean, or flat, space, the shortest path between two points is a straight line. However, in the presence of gravity, the path taken by light is not a straight line. Thus, Einstein concluded, space is *not* Euclidean.

The presence of mass, in Einstein's view, causes space to be curved. In fact, in the fully developed theory of general relativity, mass and energy (which Newton thought of as two separate entities) are interchangeable, via the famous equation $E = mc^2$. Moreover, space and time (which Newton thought of as two separate entities) form a four-dimensional space-time. A more accurate summary of Einstein's viewpoint, then, is that the presence of mass-energy causes space-time to

[2]More generally, Fermat's principle requires that the travel time be an extremum—either a minimum or a maximum. In most situations, however, the path taken by light minimizes the travel time rather than maximizing it.

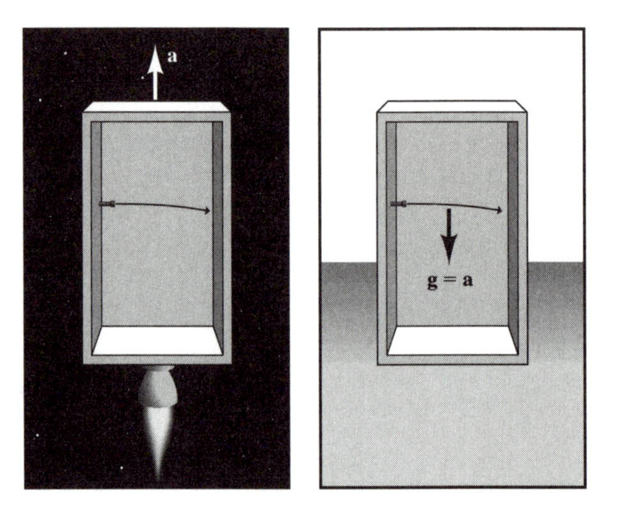

FIGURE 3.2 Equivalence principle (photon version). The path followed by a light beam in an accelerated box (left) is identical to the path followed by a light beam subjected to gravitational acceleration (right).

be curved. We now have a third way of thinking about the motion of the teddy bear in the box: (3) No forces are acting on the bear; it is simply following a *geodesic* in curved space-time.[3]

We now have two ways of describing how gravity works.

The Way of Newton:
Mass tells gravity how to exert a force ($F = -GMm/r^2$),
Force tells mass how to accelerate ($F = ma$).

The Way of Einstein:
Mass-energy tells space-time how to curve,
Curved space-time tells mass-energy how to move.[4]

Einstein's description of gravity gives a natural explanation for the equivalence principle. In the Newtonian description of gravity, the equality of the gravitational mass and the inertial mass is a remarkable coincidence. However, in Einstein's theory of general relativity, curvature is a property of space-time itself. It then follows automatically that the gravitational acceleration of an object should be independent of mass and composition—it's just following a geodesic, which is dictated by the geometry of space-time.

[3]In this context, the word "geodesic" is simply a shorter way of saying "the shortest distance between two points."

[4]This pocket summary of general relativity was coined by the physicist John Wheeler, who also popularized the term "black hole."

3.2 ■ DESCRIBING CURVATURE

In developing a mathematical theory of general relativity, in which space-time curvature is related to the mass-energy density, Einstein needed a way of mathematically describing curvature. Since picturing the curvature of a four-dimensional space-time is to say the least difficult, let's start by considering ways of describing the curvature of two-dimensional spaces, and then extend what we have learned to higher dimensions.

The simplest of two-dimensional spaces is a plane, on which Euclidean geometry holds (Figure 3.3). On a plane, a geodesic is a straight line. If a triangle is constructed on a plane by connecting three points with geodesics, the angles at its vertices (α, β, and γ in Figure 3.3) obey the relation

$$\alpha + \beta + \gamma = \pi, \tag{3.5}$$

where angles are measured in radians. On a plane, we can set up a Cartesian coordinate system, and assign to every point a coordinate (x, y). On a plane, the Pythagorean theorem holds, so the distance ds between points (x, y) and $(x + dx, y + dy)$ is given by the relation[5]

$$ds^2 = dx^2 + dy^2. \tag{3.6}$$

Stating that equation (3.6) holds true everywhere in two-dimensional space is equivalent to saying that the space is a plane. Of course, other coordinate systems can be used, in place of Cartesian coordinates. For instance, in a polar coordinate system, the distance between points (r, θ) and $(r + dr, \theta + d\theta)$ is

$$ds^2 = dr^2 + r^2 d\theta^2. \tag{3.7}$$

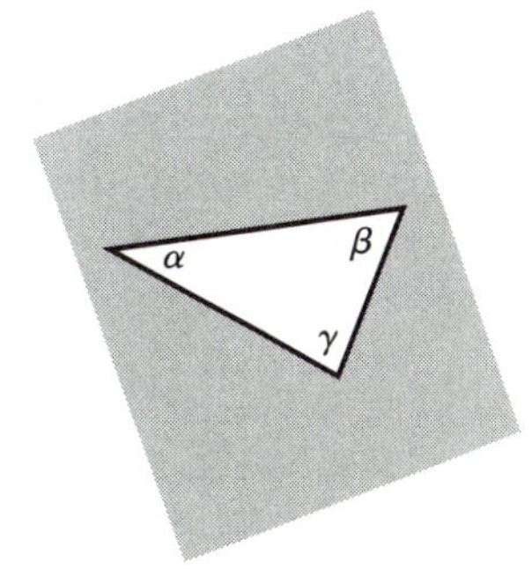

FIGURE 3.3 A flat two-dimensional space.

[5]Starting with this equation, we are adopting the convention, commonly used among relativists, that $ds^2 = (ds)^2$, and not $d(s^2)$. Omitting the parentheses simply makes the equations less cluttered.

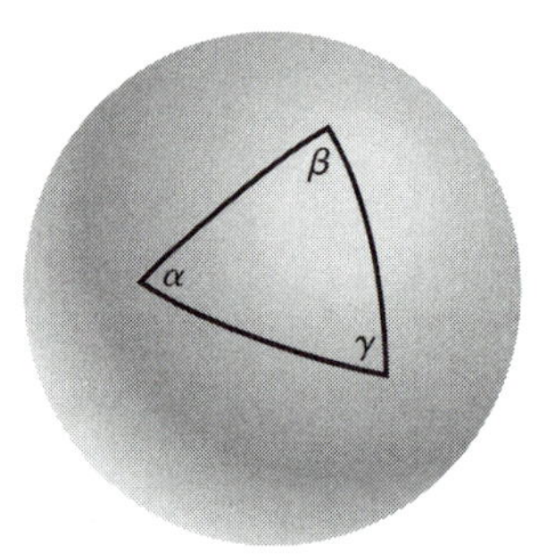

FIGURE 3.4 A positively curved two-dimensional space.

Although equations (3.6) and (3.7) are different in appearance, they both represent the same flat geometry, as you can verify by making the simple coordinate substitution $x = r\cos\theta$, $y = r\sin\theta$.

Now consider another simple two-dimensional space, the surface of a sphere (Figure 3.4). On the surface of a sphere, a geodesic is a portion of a great circle—that is, a circle whose center corresponds to the center of the sphere. If a triangle is constructed on the surface of the sphere by connecting three points with geodesics, the angles at its vertices (α, β, and γ) obey the relation

$$\alpha + \beta + \gamma = \pi + A/R^2, \tag{3.8}$$

where A is the area of the triangle, and R is the radius of the sphere. All spaces in which $\alpha + \beta + \gamma > \pi$ are called "positively curved" spaces. The surface of a sphere is a positively curved two-dimensional space. Moreover, it is a space where the curvature is homogeneous and isotropic; no matter where you draw a triangle on the surface of a sphere, or how you orient it, it must always satisfy equation (3.8).

On the surface of a sphere, we can set up a polar coordinate system by picking a pair of antipodal points to be the "north pole" and "south pole" and by picking a geodesic from the north to south pole to be the "prime meridian." If r is the distance from the north pole, and θ is the azimuthal angle measured relative to the prime meridian, then the distance ds between a point (r, θ) and another nearby point $(r + dr, \theta + d\theta)$ is given by the relation

$$ds^2 = dr^2 + R^2 \sin^2(r/R)d\theta^2. \tag{3.9}$$

Note that the surface of a sphere has a finite area, equal to $4\pi R^2$, and a maximum possible distance between points; the distance between antipodal points, at the maximum possible separation, is πR. By contrast, a plane has infinite area, and has no upper limits on the possible distance between points.[6]

[6]Since the publishers objected to producing a book of infinite size, Figure 3.3 actually shows only a portion of a plane.

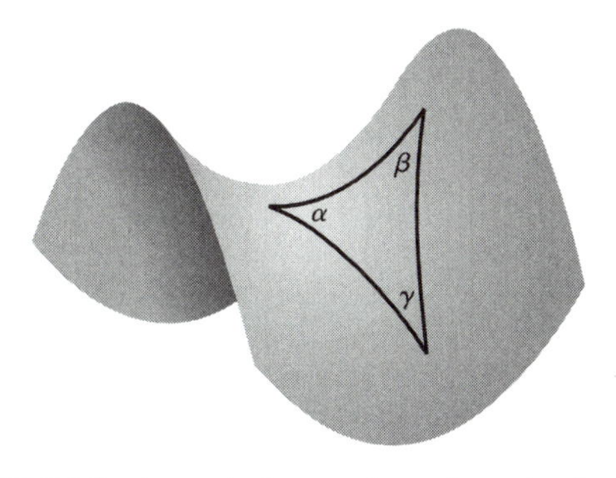

FIGURE 3.5 A negatively curved two-dimensional space.

In addition to flat spaces and positively curved spaces, there exist negatively curved spaces. An example of a negatively curved two-dimensional space is the hyperboloid, or saddle shape, shown in Figure 3.5. For illustrative purposes, we will show you a surface of *constant* negative curvature, just as the surface of a sphere has constant positive curvature.[7] Unfortunately, the mathematician David Hilbert proved that a two-dimensional surface of constant negative curvature cannot be constructed in a three-dimensional Euclidean space. The saddle shape illustrated in Figure 3.5 has constant curvature only in the central region, near the "seat" of the saddle.

Despite the difficulties in visualizing a surface of constant negative curvature, its properties can be written down easily. Consider a two-dimensional surface of constant negative curvature, with radius of curvature R. If a triangle is constructed on this surface by connecting three points with geodesics, the angles at its vertices (α, β, and γ) obey the relation

$$\alpha + \beta + \gamma = \pi - A/R^2, \tag{3.10}$$

where A is the area of the triangle.

On a surface of constant negative curvature, we can set up a polar coordinate system by choosing some point as the pole, and some geodesic leading away from the pole as the prime meridian. If r is the distance from the pole, and θ is the azimuthal angle measured relative to the prime meridian, then the distance ds between a point (r, θ) and a nearby point $(r + dr, \theta + d\theta)$ is given by

$$ds^2 = dr^2 + R^2 \sinh^2(r/R) d\theta^2. \tag{3.11}$$

A surface of constant negative curvature has infinite area, and has no upper limit on the possible distance between points.

Relations like those presented in equations (3.7), (3.9), and (3.11), which give the distance ds between two nearby points in space, are known as *metrics*. In

[7]A space with constant curvature is one where the curvature is homogeneous and isotropic.

general, curvature is a local property. A tablecloth can be badly rumpled at one end of the table and smooth at the other end; a bagel (or other toroidal object) is negatively curved on part of its surface and positively curved on other portions.[8] However, if you want a two-dimensional space to be homogeneous and isotropic, only three possibilities can fit the bill: the space can be uniformly *flat*; it can have uniform *positive* curvature; or it can have uniform *negative* curvature. Thus, if a two-dimensional space has curvature that is homogeneous and isotropic, its geometry can be specified by two quantities, κ, and R. The number κ, called the *curvature constant*, is $\kappa = 0$ for a flat space, $\kappa = +1$ for a positively curved space, and $\kappa = -1$ for a negatively curved space. If the space is curved, then the quantity R, which has dimensions of length, is the radius of curvature.

The results for two-dimensional space can be extended straightforwardly to three dimensions. A three-dimensional space, if its curvature is homogeneous and isotropic, must be flat, or have uniform positive curvature, or have uniform negative curvature. If a three-dimensional space is flat ($\kappa = 0$), it has the metric

$$ds^2 = dx^2 + dy^2 + dz^2, \tag{3.12}$$

expressed in Cartesian coordinates, or

$$ds^2 = dr^2 + r^2[d\theta^2 + \sin^2\theta d\phi^2], \tag{3.13}$$

expressed in spherical coordinates.

If a three-dimensional space has uniform positive curvature ($\kappa = +1$), its metric is

$$ds^2 = dr^2 + R^2 \sin^2(r/R)[d\theta^2 + \sin^2\theta d\phi^2]. \tag{3.14}$$

A positively curved three-dimensional space has finite volume, just as a positively curved two-dimensional space has finite area. The point at $r = \pi R$ is the antipodal point to the origin, just as the south pole, at $r = \pi R$, is the antipodal point to the north pole, at $r = 0$, on the surface of a sphere. By traveling a distance $C = 2\pi R$, it is possible to "circumnavigate" a space of uniform positive curvature.

Finally, if a three-dimensional space has uniform negative curvature ($\kappa = -1$), its metric is

$$ds^2 = dr^2 + R^2 \sinh^2(r/R)[d\theta^2 + \sin^2\theta d\phi^2]. \tag{3.15}$$

Like flat space, negatively curved space has infinite volume.

The three possible metrics for a homogeneous, isotropic, three-dimensional space can be written more compactly in the form

$$ds^2 = dr^2 + S_\kappa(r)^2 d\Omega^2, \tag{3.16}$$

[8] You can test this assertion, if you like, by drawing triangles on a bagel.

where

$$d\Omega^2 \equiv d\theta^2 + \sin^2\theta d\phi^2 \tag{3.17}$$

and

$$S_\kappa(r) = \begin{cases} R\sin(r/R) & (\kappa = +1) \\ r & (\kappa = 0) \\ R\sinh(r/R) & (\kappa = -1). \end{cases} \tag{3.18}$$

Note that in the limit $r \ll R$, $S_\kappa \approx r$, regardless of the value of κ. When space is flat, or negatively curved, S_κ increases monotonically with r, with $S_\kappa \to \infty$ as $r \to \infty$. By contrast, when space is positively curved, S_κ increases to a maximum of $S_{\rm max} = R$ at $r/R = \pi/2$, then decreases again to 0 at $r/R = \pi$, the antipodal point to the origin.

The coordinate system (r, θ, ϕ) is not the only possible system. For instance, if we switch the radial coordinate from r to $x \equiv S_\kappa(r)$, the metric for a homogeneous, isotropic, three-dimensional space can be written in the form

$$ds^2 = \frac{dx^2}{1 - \kappa x^2/R^2} + x^2 d\Omega^2. \tag{3.19}$$

Although the metrics written in equations (3.16) and (3.19) appear different on the page, they represent the same homogeneous, isotropic spaces. They merely have a different functional form because of the different choice of radial coordinates.

3.3 ■ THE ROBERTSON–WALKER METRIC

So far, we've only considered the metrics for simple two-dimensional and three-dimensional spaces. However, relativity teaches us that space and time together comprise a four-dimensional space-time. Just as we can compute the distance between two points in space using the appropriate metric for that space, so we can compute the four-dimensional distance between two events in space-time. Consider two events, one occurring at the space-time location (t, r, θ, ϕ), and another occurring at the space-time location $(t + dt, r + dr, \theta + d\theta, \phi + d\phi)$. According to the laws of special relativity, the space-time separation between these two events is

$$ds^2 = -c^2 dt^2 + dr^2 + r^2 d\Omega^2. \tag{3.20}$$

The metric given in equation (3.20) is called the *Minkowski metric*, and the space-time that it describes is called Minkowski space-time. Note, from a comparison with equation (3.16), that the spatial component of Minkowski space-time is Euclidean, or flat.

A photon's path through space-time is a four-dimensional geodesic—and not just any geodesic, mind you, but a special variety called a *null geodesic*. A null geodesic is one for which, along every infinitesimal segment of the photon's path, $ds = 0$. In Minkowski space-time, then, a photon's trajectory obeys the relation

$$ds^2 = 0 = -c^2dt^2 + dr^2 + r^2d\Omega^2. \tag{3.21}$$

If the photon is moving along a radial path, toward or away from the origin, this means, since θ and ϕ are constant,

$$c^2dt^2 = dr^2, \tag{3.22}$$

or

$$\frac{dr}{dt} = \pm c. \tag{3.23}$$

The Minkowski metric of equation (3.20) applies only within the context of special relativity, so called because it deals with the special case in which space-time is not curved by the presence of mass and energy. Without any gravitational effects, Minkowski space-time is flat and static. When gravity is added, however, the permissible space-times are more interesting. In the 1930s, the physicists Howard Robertson and Arthur Walker asked, "What form can the metric of space-time assume if the universe is spatially homogeneous and isotropic at all time, and if distances are allowed to expand (or contract) as a function of time?" The metric they derived (independently of each other) is called the *Robertson–Walker metric*. It is most generally written in the form

$$ds^2 = -c^2dt^2 + a(t)^2\left[\frac{dx^2}{1-\kappa x^2/R_0^2} + x^2d\Omega^2\right]. \tag{3.24}$$

Note that the spatial component of the Robertson–Walker metric consists of the spatial metric for a uniformly curved space of radius R_0 (compare equation (3.19)), scaled by the square of the scale factor $a(t)$. The scale factor, first introduced in section 2.3, describes how distances in a homogeneous, isotropic universe expand or contract with time. The Robertson–Walker metric can also be written in the form (see equation (3.16))

$$ds^2 = -c^2dt^2 + a(t)^2\left[dr^2 + S_\kappa(r)^2d\Omega^2\right], \tag{3.25}$$

with the function $S_\kappa(r)$ for the three different types of curvature given by equation (3.18).

The time variable t in the Robertson–Walker metric is the cosmological proper time, called the *cosmic time* for short, and is the time measured by an observer who sees the universe expanding uniformly around him. The spatial variables (x, θ, ϕ) or (r, θ, ϕ) are called the *comoving coordinates* of a point in space; if the expansion of the universe is perfectly homogeneous and isotropic, then the comoving coordinates of any point remain constant with time.

The assumption of homogeneity and isotropy is a very powerful one. If the universe is perfectly homogeneous and isotropic, then everything we need to know about its geometry is contained within $a(t)$, κ, and R_0. The scale factor $a(t)$ is a dimensionless function of time that describes how distances grow or decrease with time; it is normalized so that $a(t_0) = 1$ at the present moment. The curvature constant κ is a dimensionless number that can take on one of three discrete values: $\kappa = 0$ if the universe is spatially flat, $\kappa = -1$ if the universe has negative spatial curvature, and $\kappa = +1$ if the universe has positive spatial curvature. The radius of curvature R_0 has dimensions of length, and gives the radius of curvature of the universe at the present moment. Much of modern cosmology, as we'll see in later chapters, is devoted in one way or another to finding the values of $a(t)$, κ, and R_0. The assumption of spatial homogeneity and isotropy is so powerful, Robertson and Walker assumed it in the 1930s, long before the available observational evidence gave any support for such an assumption. If homogeneity and isotropy did not exist, as Voltaire might have said, it would be necessary to invent them—at least if your desire is to have a simple, analytically tractable form for the metric of space-time.

In truth, the observations reveal that the universe is *not* homogeneous and isotropic on small scales. Thus, the Robertson–Walker metric is only an approximation that holds good on large scales. On smaller scales, the universe is "lumpy," and hence does not expand uniformly. Small, dense lumps, such as humans, teddy bears, and interstellar dust grains, are held together by electromagnetic forces, and hence do not expand. Larger lumps, as long as they are sufficiently dense, are held together by their own gravity, and hence do not expand. Examples of such gravitationally bound systems are planetary systems (such as the solar system in which we live), galaxies (such as the galaxy in which we live), and clusters of galaxies (such as the Local Group in which we live). It's only on scales larger than $\sim 100\,\text{Mpc}$ that the expansion of the universe can be treated as the ideal, homogeneous, isotropic expansion described by the single scale factor $a(t)$.

3.4 ■ PROPER DISTANCE

Consider a galaxy far away from us—sufficiently far away that we may ignore the small scale perturbations of space-time and adopt the Robertson–Walker metric. One question we may ask is, "Exactly how far away is this galaxy?" In an expanding universe, the distance between two objects is increasing with time. Thus, if we want to assign a spatial distance d between two objects, we must specify the time t at which the distance is the correct one. Suppose that you are at the origin, and that the galaxy that you are observing is at a comoving coordinate position (r, θ, ϕ), as illustrated in Figure 3.6. The proper distance $d_p(t)$ between two points is equal to the length of the spatial geodesic between them when the scale factor is fixed at the value $a(t)$. The proper distance between the observer and galaxy in Figure 3.6 can be found using the Robertson–Walker metric at a fixed time t:

$$ds^2 = a(t)^2[dr^2 + S_\kappa(r)^2 d\Omega^2]. \tag{3.26}$$

FIGURE 3.6 An observer at the origin observes a galaxy at coordinate position (r, θ, ϕ). A photon emitted by the galaxy at cosmic time t_e reaches the observer at cosmic time t_0.

Along the spatial geodesic between the observer and galaxy, the angle (θ, ϕ) is constant, and thus

$$ds = a(t)dr. \tag{3.27}$$

The proper distance d_p is found by integrating over the radial comoving coordinate r:

$$d_p(t) = a(t) \int_0^r dr = a(t)r. \tag{3.28}$$

Alternatively, if you wish to use the spatial coordinates (x, θ, ϕ) instead of (r, θ, ϕ), where $x = S_\kappa(r)$, you may invert the relations of equation (3.18) to find

$$d_p(t) = a(t)r(x) = \begin{cases} a(t)R_0 \sin^{-1}(x/R_0) & (\kappa = +1) \\ a(t)x & (\kappa = 0) \\ a(t)R_0 \sinh^{-1}(x/R_0) & (\kappa = -1). \end{cases} \tag{3.29}$$

Because the proper distance has the form $d_p(t) = a(t)r$, with the comoving coordinate r constant with time, the rate of change for the proper distance between us and a distant galaxy is

$$\dot{d}_p = \dot{a}r = \frac{\dot{a}}{a}d_p. \tag{3.30}$$

Thus, at the current time $(t = t_0)$, there is a linear relation between the proper distance to a galaxy and its recession speed:

$$v_p(t_0) = H_0 d_p(t_0), \tag{3.31}$$

where

$$v_p(t_0) \equiv \dot{d}_p(t_0) \tag{3.32}$$

and

$$H_0 = \left(\frac{\dot{a}}{a}\right)_{t=t_0} . \tag{3.33}$$

In a sense, this is just a repetition of what was demonstrated in section 2.3; if the distance between points is proportional to $a(t)$, there will be a linear relation between the relative velocity of two points and the distance between them. Now, however, we are interpreting the change in distance between widely separated galaxies as being associated with the expansion of space. As the distance between galaxies increases, the radius of curvature of the universe, $R(t) = a(t)R_0$, increases at the same rate.

Some cosmology books will contain a statement like "As space expands, it drags galaxies away from each other." Statements of this sort are misleading because they make galaxies appear to be entirely passive. On the other hand, a statement like "As galaxies move apart, they drag space along with them" would be equally misleading because it makes space appear to be entirely passive. As the theory of general relativity points out, space-time and mass-energy are intimately linked. Yes, the curvature of space-time does tell mass-energy how to move, but then it's mass-energy which tells space-time how to curve.

The linear velocity-distance relation given in equation (3.31) implies that points separated by a proper distance greater than a critical value

$$d_H(t_0) \equiv c/H_0, \tag{3.34}$$

generally called the *Hubble distance*, will have

$$v_p = \dot{d}_p > c. \tag{3.35}$$

Using the observationally determined value of $H_0 = 70 \pm 7\,\text{km}\,\text{s}^{-1}\,\text{Mpc}^{-1}$, the current value of the Hubble distance in our universe is

$$d_H(t_0) = 4300 \pm 400\,\text{Mpc}. \tag{3.36}$$

Thus, galaxies farther than 4300 megaparsecs from us are currently moving away from us at speeds greater than that of light. Cosmological innocents sometimes exclaim, "Gosh! Doesn't this violate the law that massive objects can't travel faster than the speed of light?" Actually, it doesn't. The speed limit that states that massive objects must travel with $v < c$ relative to each other is one of the results of special relativity, and refers to the relative motion of objects within a static space. In the context of general relativity, there is no objection to having two points moving away from each other at superluminal speed due to the expansion of space.

When we observe a distant galaxy, we know its angular position very well, but not its distance. That is, we can point in its direction, but we don't know its current proper distance $d_p(t_0)$. We can, however, measure the redshift z of the light we receive from the galaxy. Although the redshift doesn't tell us the proper distance to the galaxy, it does tell us what the scale factor a was at the time the light from that galaxy was emitted. To see the link between a and z, consider the galaxy illustrated in Figure 3.6. Light that was emitted by the galaxy at a time t_e is observed by us at a time t_0. During its travel from the distant galaxy to us, the light traveled along a null geodesic, with $ds = 0$. The null geodesic has θ and ϕ constant.[9] Thus, along the light's null geodesic,

$$c^2dt^2 = a(t)^2dr^2. \tag{3.37}$$

Rearranging this relation, we find

$$c\frac{dt}{a(t)} = dr. \tag{3.38}$$

In equation (3.38), the left-hand side is a function only of t, and the right-hand side is independent of t. Suppose the distant galaxy emits light with a wavelength λ_e, as measured by an observer in the emitting galaxy. Fix your attention on a single wave crest of the emitted light. The wave crest is emitted at a time t_e and observed at a time t_0, such that

$$c\int_{t_e}^{t_0}\frac{dt}{a(t)} = \int_0^r dr = r. \tag{3.39}$$

The next wave crest of light is emitted at a time $t_e + \lambda_e/c$, and is observed at a time $t_0 + \lambda_0/c$, where, in general, $\lambda_0 \neq \lambda_e$. For the second wave crest,

$$c\int_{t_e+\lambda_e/c}^{t_0+\lambda_0/c}\frac{dt}{a(t)} = \int_0^r dr = r. \tag{3.40}$$

Comparing equations (3.39) and (3.40), we find that

$$\int_{t_e}^{t_0}\frac{dt}{a(t)} = \int_{t_e+\lambda_e/c}^{t_0+\lambda_0/c}\frac{dt}{a(t)}. \tag{3.41}$$

That is, the integral of $dt/a(t)$ between the time of emission and the time of observation is the same for every wave crest in the emitted light. If we subtract the integral

$$\int_{t_e+\lambda_e/c}^{t_0}\frac{dt}{a(t)} \tag{3.42}$$

[9] In a homogeneous, isotropic universe there's no reason for the light to swerve to one side or the other.

from each side of equation (3.41), we find the relation

$$\int_{t_e}^{t_e+\lambda_e/c} \frac{dt}{a(t)} = \int_{t_0}^{t_0+\lambda_0/c} \frac{dt}{a(t)}. \tag{3.43}$$

That is, the integral of $dt/a(t)$ between the emission of successive wave crests is equal to the integral of $dt/a(t)$ between the observation of successive wave crests. This relation becomes still simpler when we realize that during the time between the emission or observation of two wave crests, the universe doesn't have time to expand by a significant amount. The time scale for expansion of the universe is the Hubble time, $H_0^{-1} \approx 14\,\text{Gyr}$. The time between wave crests, for visible light, is $\lambda/c \approx 2 \times 10^{-15}\,\text{s} \approx 10^{-32} H_0^{-1}$. Thus, $a(t)$ is effectively constant in the integrals of equation (3.43). Thus, we may write

$$\frac{1}{a(t_e)} \int_{t_e}^{t_e+\lambda_e/c} dt = \frac{1}{a(t_0)} \int_{t_0}^{t_0+\lambda_0/c} dt, \tag{3.44}$$

or

$$\frac{\lambda_e}{a(t_e)} = \frac{\lambda_0}{a(t_0)}. \tag{3.45}$$

Using the definition of redshift, $z = (\lambda_0 - \lambda_e)/\lambda_e$, we find that the redshift of light from a distant object is related to the expansion factor at the time it was emitted via the equation

$$1 + z = \frac{a(t_0)}{a(t_e)} = \frac{1}{a(t_e)}. \tag{3.46}$$

Here, we have used the usual convention that $a(t_0) = 1$.

Thus, if we observe a galaxy with a redshift $z = 2$, we are observing it as it was when the universe had a scale factor $a(t_e) = \frac{1}{3}$. The redshift we observe for a distant object depends only on the relative scale factors at the time of emission and the time of observation. It doesn't depend on how the transition between $a(t_e)$ and $a(t_0)$ was made. It doesn't matter if the expansion was gradual or abrupt; it doesn't matter if the transition was monotonic or oscillatory. All that matters is the scale factors at the time of emission and the time of observation.

SUGGESTED READING

Full references are given in the Annotated Bibliography on page 235.

Harrison (2000), ch. 10–12: Curved space and relativity (both special and general)

Narlikar (2002), ch. 2–3: Delves deeper into general relativity while discussing the Robertson–Walker metric

Peacock (1999), ch. 3.1: Derivation and discussion of the Robertson–Walker metric

Rich (2001), ch. 3: Coordinates and metrics in the context of general relativity

PROBLEMS

3.1. What evidence can you provide to support the assertion that the universe is electrically neutral on large scales?

3.2. Suppose you are a two-dimensional being, living on the surface of a sphere with radius R. An object of width $ds \ll R$ is at a distance r from you (remember, all distances are measured on the surface of the sphere). What angular width $d\theta$ will you measure for the object? Explain the behavior of $d\theta$ as $r \rightarrow \pi R$.

3.3. Suppose you are *still* a two-dimensional being, living on the same sphere of radius R. Show that if you draw a circle of radius r, the circle's circumference will be

$$C = 2\pi R \sin(r/R). \tag{3.47}$$

Idealize the Earth as a perfect sphere of radius $R = 6371\,\text{km}$. If you could measure distances with an error of ± 1 meter, how large a circle would you have to draw on the Earth's surface to convince yourself that the Earth is spherical rather than flat?

3.4. Consider an equilateral triangle, with sides of length L, drawn on a two-dimensional surface of constant curvature. Can you draw an equilateral triangle of arbitrarily large area A on a surface with $\kappa = +1$ and radius of curvature R? If not, what is the maximum possible value of A? Can you draw an equilateral triangle of arbitrarily large area A on a surface with $\kappa = 0$? If not, what is the maximum possible value of A? Can you draw an equilateral triangle of arbitrarily large area A on a surface with $\kappa = -1$ and radius of curvature R? If not, what is the maximum possible value of A?

3.5. By making the substitutions $x = r \sin\theta \cos\phi$, $y = r \sin\theta \sin\phi$, and $z = r \cos\theta$, demonstrate that equations (3.12) and (3.13) represent the same metric.

CHAPTER 4

Cosmic Dynamics

In a homogeneous and isotropic universe, but one that is allowed to expand or contract with time, everything you need to know about the curvature is given by κ, R_0, and $a(t)$. The curvature constant κ gives the sign of the curvature: positive ($\kappa = +1$), negative ($\kappa = -1$), or flat ($\kappa = 0$). If κ is nonzero, then R_0 is the radius of curvature of the universe, as measured at the present moment ($t = t_0$). Finally, the scale factor $a(t)$ tells how distances in the universe increase with time as the universe expands, or decrease with time as the universe contracts. The scale factor is normalized so that $a(t_0) = 1$ at the present moment.

The idea that the universe could be curved, or non-Euclidean, actually long predates Einstein's theory of general relativity. As early as 1829, half a century before Einstein's birth, Nikolai Ivanovich Lobachevski, one of the founders of non-Euclidean geometry, proposed observational tests to demonstrate whether the universe was curved. In principle, measuring the curvature of the universe is simple; in practice it is much more difficult. In principle, we could determine the curvature by drawing a really, really big triangle, and measuring the angles α, β, and γ at the vertices. Equations (3.5), (3.8), and (3.10) generalize to the equation

$$\alpha + \beta + \gamma = \pi + \frac{\kappa A}{R_0^2}, \tag{4.1}$$

where A is the area of the triangle. Therefore, if $\alpha + \beta + \gamma > \pi$ radians, the universe is positively curved, and if $\alpha + \beta + \gamma < \pi$ radians, the universe is negatively curved. If, in addition, we measure the area of the triangle, we can determine the radius of curvature R_0. Unfortunately for this elegant geometric plan, the area of the biggest triangle we can draw is much smaller than R_0^2, and the deviation of $\alpha + \beta + \gamma$ from π radians would be too small to measure.

About all we can conclude from geometric arguments is that if the universe is positively curved, it can't have a radius of curvature R_0 that is significantly smaller than the current Hubble distance, $c/H_0 \approx 4300\,\text{Mpc}$. To understand why this is so, recall that if our universe is positively curved, it has finite size, with a circumference currently equal to $C_0 = 2\pi R_0$. In the past, since our universe is expanding, its circumference was even smaller. Thus, if the current circumference C_0 is less than ct_0, then photons will have had time to circumnavigate the universe. If $C_0 \ll ct_0 \sim c/H_0$, then photons will have had time to circumnavigate the

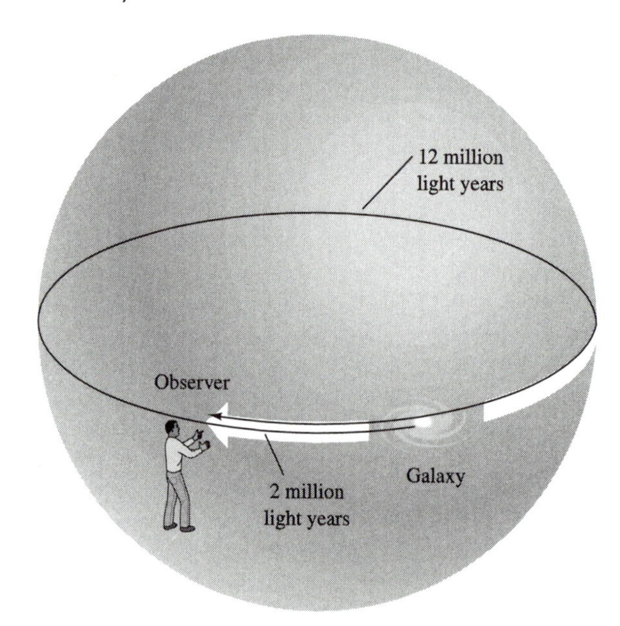

FIGURE 4.1 A two-dimensional positively curved universe, demonstrating how an observer in such a universe could see multiple images of the same galaxy.

universe many times. To take an extreme example, suppose the universe were positively curved with a circumference of only 10 million light years (roughly 3 Mpc). The two-dimensional analog to such a universe is shown in Figure 4.1. Looking toward the galaxy M31, which is 2 million light years away from us, we would see one image of M31, comprised of photons that had traveled 2 million light years, showing M31 as it was 2 million years ago. We would also see another image, comprised of photons that had traveled 12 million light years, showing M31 as it was 12 million years ago. And so on. Moreover, looking in the exact opposite direction to M31, we would see an image of M31, comprised of photons that had traveled 8 million light years, showing M31 as it was 8 million years ago. We would also see another image, comprised of photons that had traveled 18 million light years, showing M31 as it was 18 million years ago.[1] And so on. Since we don't see periodicities of this sort, we conclude that if the universe is positively curved, its radius of curvature R_0 must be very large—comparable to or larger than the current Hubble distance c/H_0.

4.1 ■ THE FRIEDMANN EQUATION

Although 19th century mathematicians and physicists such as Lobachevski were able to conceive of curved space, it wasn't until Albert Einstein first published his

[1]This assumes that $2\pi R_0/c \ll H_0^{-1}$, and that the universe therefore doesn't expand significantly as a photon goes once or twice around the universe.

theory of general relativity in 1915 that the curvature of space-time was linked to its mass-energy content. The key equation of general relativity is Einstein's field equation, which is the relativistic equivalent of Poisson's equation in Newtonian dynamics. Poisson's equation,

$$\nabla^2 \Phi = 4\pi G\rho, \tag{4.2}$$

gives a mathematical relation between the gravitational potential Φ at a point in space and the mass density ρ at that point. By taking the gradient of the potential, you determine the acceleration, and then can compute the trajectory of objects moving freely through space. Einstein's field equation, by contrast, gives a mathematical relation between the metric of space-time at a point and the energy and pressure at that space-time point. The trajectories of freely moving objects then correspond to geodesics in curved space-time.

In a cosmological context, Einstein's field equations can be used to find the linkage between $a(t)$, κ, and R_0, which describe the curvature of the universe, and the energy density $\varepsilon(t)$ and pressure $P(t)$ of the contents of the universe. The equation that links together $a(t)$, κ, R_0, and $\varepsilon(t)$ is known as the *Friedmann equation*, after Alexander Alexandrovich Friedmann, who first derived the equation in 1922. Friedmann actually started his scientific career as a meteorologist. Later, however, he taught himself general relativity, and used Einstein's field equations to describe how a spatially homogeneous and isotropic universe expands or contracts as a function of time. It is intriguing to note that Friedmann published his first results, implying an expanding or contracting universe, seven years before Hubble published Hubble's Law in 1929. Unfortunately, Friedmann's papers received little notice at first. Even Einstein initially dismissed Friedmann's work as a mathematical curiosity, unrelated to the universe we actually live in. It wasn't until Hubble's results were published that Einstein acknowledged the reality of the expanding universe. Alas, Friedmann did not live to see his vindication; he died of typhoid fever in 1925, when he was only 37 years old.

Friedmann derived his eponymous equation starting from Einstein's field equation, using the full power of general relativity. Even without bringing relativity into play, some (though not all) of the aspects of the Friedmann equation can be understood with the use of purely Newtonian dynamics. To see how the expansion or contraction of the universe can be viewed from a Newtonian viewpoint, we will first derive the nonrelativistic equivalent of the Friedmann equation, starting from Newton's Law of Gravity and Second Law of Motion. Then we will state (without proof) the modifications that must be made to find the more correct, general relativistic form of the Friedmann equation.

To begin, consider a homogeneous sphere of matter, with total mass M_s constant with time (Figure 4.2). The sphere is expanding or contracting isotropically, so that its radius $R_s(t)$ is increasing or decreasing with time. Place a test mass, of infinitesimal mass m, at the surface of the sphere. The gravitational force F

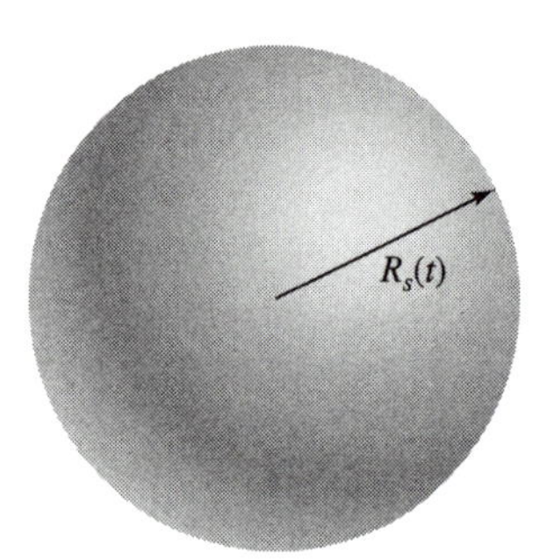

FIGURE 4.2 A sphere of radius $R_s(t)$ and mass M_s, expanding or contracting under its own gravity.

experienced by the test mass will be, from Newton's Law of Gravity,

$$F = -\frac{GM_s m}{R_s(t)^2}. \tag{4.3}$$

The gravitational acceleration at the surface of the sphere will then be, from Newton's Second Law of Motion,

$$\frac{d^2 R_s}{dt^2} = -\frac{GM_s}{R_s(t)^2}. \tag{4.4}$$

Multiply each side of the equation by dR_s/dt and integrate to find

$$\frac{1}{2}\left(\frac{dR_s}{dt}\right)^2 = \frac{GM_s}{R_s(t)} + U, \tag{4.5}$$

where U is a constant of integration. Equation (4.5) simply states that the sum of the *kinetic* energy per unit mass,

$$E_{\text{kin}} = \frac{1}{2}\left(\frac{dR_s}{dt}\right)^2, \tag{4.6}$$

and the gravitational *potential* energy per unit mass,

$$E_{\text{pot}} = -\frac{GM_s}{R_s(t)}, \tag{4.7}$$

is constant for a bit of matter at the surface of a sphere, as the sphere expands or contracts under its own gravitational influence.

Since the mass of the sphere is constant as it expands or contracts, we may write

$$M_s = \frac{4\pi}{3}\rho(t)R_s(t)^3. \tag{4.8}$$

Since the expansion is isotropic about the sphere's center, we may write the radius $R_s(t)$ in the form

$$R_s(t) = a(t)r_s, \tag{4.9}$$

where $a(t)$ is the scale factor and r_s is the comoving radius of the sphere. In terms of $\rho(t)$ and $a(t)$, the energy conservation equation (4.5) can be rewritten in the form

$$\frac{1}{2}r_s^2\dot{a}^2 = \frac{4\pi}{3}Gr_s^2\rho(t)a(t)^2 + U. \tag{4.10}$$

Dividing each side of equation (4.10) by $r_s^2a^2/2$ yields the equation

$$\left(\frac{\dot{a}}{a}\right)^2 = \frac{8\pi G}{3}\rho(t) + \frac{2U}{r_s^2}\frac{1}{a(t)^2}. \tag{4.11}$$

Equation (4.11) gives the *Friedmann equation* in its Newtonian form.

Note that the time derivative of the scale factor only enters into equation (4.11) as $\dot{a}^2$; a contracting sphere ($\dot{a} < 0$) is simply the time reversal of an expanding sphere ($\dot{a} > 0$). Let's concentrate on the case of an expanding sphere, analogous to the expanding universe in which we find ourselves. The future of the expanding sphere falls into one of three classes, depending on the sign of U. First, consider the case $U > 0$. In this case, the right-hand side of equation (4.11) is always positive. Therefore, $\dot{a}^2$ is always positive, and the expansion of the sphere never stops. Second, consider the case $U < 0$. In this case, the right-hand side of equation (4.11) starts out positive. However, at a maximum scale factor

$$a_{\max} = -\frac{GM_s}{Ur_s}, \tag{4.12}$$

the right-hand side will equal zero, and expansion will stop. Since $\ddot{a}$ will still be negative, the sphere will then contract. Third, and finally, consider the case $U = 0$. This is the boundary case in which $\dot{a} \to 0$ as $t \to \infty$ and $\rho \to 0$.

The three possible fates of an expanding sphere in a Newtonian universe are analogous to the three possible fates of a ball thrown upward from the surface of the Earth. First, the ball can be thrown upward with a speed greater than the escape speed; in this case, the ball continues to go upward forever. Second, the ball can be thrown upward with a speed less than the escape speed; in this case, the ball reaches a maximum altitude, then falls back down. Third, and finally, the ball can be thrown upward with a speed exactly equal to the escape speed; in this case, the speed of the ball approaches zero as $t \to \infty$.

The Friedmann equation in its Newtonian form (equation (4.11)) is useful in picturing how isotropically expanding objects behave under the influence of their self-gravity. However, its application to the real universe must be regarded with considerable skepticism. First of all, a spherical volume of finite radius R_s cannot represent a homogeneous, isotropic universe. In a finite spherical volume, there exists a special location (the center of the sphere), violating the principle of homogeneity, and at any point there exists a special direction (the direction pointing toward the center), violating the principle of isotropy. We may instead regard the sphere of radius R_s as being carved out of an infinite, homogeneous, isotropic universe. In that case, Newtonian dynamics tell us that the gravitational acceleration inside a hollow spherically symmetric shell is equal to zero. We divide up the region outside the sphere into concentric shells, and thus conclude that the test mass m at R_s experiences no net acceleration from matter at $R > R_s$. Unfortunately, a Newtonian argument of this sort assumes that space is Euclidean. A derivation of the correct Friedmann equation, including the possibility of spatial curvature, has to begin with Einstein's field equations.

The correct form of the Friedmann equation, including all general relativistic effects, is

$$\left(\frac{\dot{a}}{a}\right)^2 = \frac{8\pi G}{3c^2}\varepsilon(t) - \frac{\kappa c^2}{R_0^2}\frac{1}{a(t)^2}. \tag{4.13}$$

Note the changes made in going from the Newtonian form of the Friedmann equation (equation (4.11)) to the correct relativistic form (equation (4.13)). The first change is that the mass density ρ has been replaced by an energy density ε divided by the square of the speed of light. One of Einstein's insights was that in determining the gravitational influence of a particle, the important quantity was not its mass m but its energy,

$$E = (m^2c^4 + p^2c^2)^{1/2}. \tag{4.14}$$

Here p is the momentum of the particle as seen by an observer at the particle's location, who sees the universe expanding isotropically around her. Any motion that a particle has, in addition to the motion associated with the expansion or contraction of the universe, is called the particle's *peculiar* motion.[2] If a massive particle is nonrelativistic—that is, if its peculiar velocity v is much less than c—then its peculiar momentum will be $p \approx mv$, and its energy will be

$$E_{\text{non-rel}} \approx mc^2(1 + v^2/c^2) \approx mc^2 + \frac{1}{2}mv^2. \tag{4.15}$$

Thus, if the universe contained only massive, slowly moving particles, then the energy density ε would be nearly equal to ρc^2, with only a small correction for the kinetic energy $mv^2/2$ of the particles. However, photons and other massless particles also have an energy,

$$E_{\text{rel}} = pc = hf, \tag{4.16}$$

[2]The adjective "peculiar" comes from the Latin "peculium," meaning "private property." The peculiar motion of a particle is thus the motion that belongs to the particle alone, and not to the global expansion or contraction of the universe.

which also contributes to the energy density ε. Not only do photons respond to the curvature of space-time, they also contribute to it.

The second change that must be made in going from the Newtonian form of the Friedmann equation to the correct relativistic form is making the substitution

$$\frac{2U}{r_s^2} = -\frac{\kappa c^2}{R_0^2}. \tag{4.17}$$

In the context of general relativity, the curvature κ is related to the Newtonian energy U of a test mass. The case with $U < 0$ corresponds to positive curvature ($\kappa = +1$), while the case with $U > 0$ corresponds to negative curvature ($\kappa = -1$). The special case with $U = 0$ corresponds to the special case where the space is perfectly flat ($\kappa = 0$). Although I have not given the derivation of the Friedmann equation in the general relativistic case, it makes sense that the curvature, given by κ and R_0, the expansion rate, given by $a(t)$, and the energy density ε should be bound up together in the same equation. After all, in Einstein's view, the energy density of the universe determines both the curvature of space and the overall dynamics of the expansion.

The Friedmann equation is a Very Important Equation in cosmology.[3] However, if we want to apply the Friedmann equation to the real universe, we must have some way of tying it to observable properties. For instance, the Friedmann equation can be tied to the Hubble constant, H_0. Remember, in a universe whose expansion (or contraction) is described by a scale factor $a(t)$, there's a linear relation between recession speed v and proper distance d:

$$v(t) = H(t)d(t), \tag{4.18}$$

where $H(t) \equiv \dot{a}/a$. Thus, the Friedmann equation can be rewritten in the form

$$H(t)^2 = \frac{8\pi G}{3c^2}\varepsilon(t) - \frac{\kappa c^2}{R_0^2 a(t)^2}. \tag{4.19}$$

At the present moment,

$$H_0 = H(t_0) = \left(\frac{\dot{a}}{a}\right)_{t=t_0} = 70 \pm 7\,\mathrm{km\,s^{-1}\,Mpc^{-1}}. \tag{4.20}$$

As an etymological aside, I should point out that the time-varying function $H(t)$ is generally known as the "Hubble parameter," while H_0, the value of $H(t)$ at the present day, is known as the "Hubble constant."

The Friedmann equation evaluated at the present moment is

$$H_0^2 = \frac{8\pi G}{3c^2}\varepsilon_0 - \frac{\kappa c^2}{R_0^2}, \tag{4.21}$$

[3] You should consider writing it in reverse on your forehead so that you can see it every morning in the mirror when you comb your hair.

using the convention that a subscript "0" indicates the value of a time-varying quantity evaluated at the present. Thus, the Friedmann equation gives a relation among H_0, which tells us the current rate of expansion, ε_0, which tells us the current energy density, and κ/R_0^2, which tells us the current curvature. Due to the difficulty of measuring the curvature directly by geometric means, it is useful to have an indirect method of determining κ and R_0. If we were able to measure H_0 and ε_0 with high precision, we could use equation (4.21) to determine the curvature. Even without knowledge of the current density ε_0, we can use equation (4.21) to place a lower limit on R_0 in a *negatively* curved universe. If we assume ε_0 is nonnegative, then for a given value of H_0, the product κ/R_0^2 is minimized in the limit $\varepsilon_0 \to 0$. In the limit of a totally empty universe, with no energy content, the curvature is negative, with a radius of curvature

$$R_0(\text{min}) = c/H_0. \tag{4.22}$$

This is the minimum radius of curvature that a negatively curved universe can have, assuming that general relativity correctly describes the curvature. Since we know that the universe contains matter and radiation, and hence that $\varepsilon_0 > 0$, the radius of curvature must be greater than the Hubble distance if the universe is negatively curved.[4]

As we have seen, the Friedmann equation can generally be written as

$$H(t)^2 = \frac{8\pi G}{3c^2}\varepsilon(t) - \frac{\kappa c^2}{R_0^2 a(t)^2}, \tag{4.23}$$

for all universes with a Robertson–Walker metric whose expansion or contraction is governed by the rules of general relativity. In a spatially flat universe ($\kappa = 0$), the Friedmann equation takes a particularly simple form:

$$H(t)^2 = \frac{8\pi G}{3c^2}\varepsilon(t). \tag{4.24}$$

Thus, for a given value of the Hubble parameter, there is a *critical density*,

$$\varepsilon_c(t) \equiv \frac{3c^2}{8\pi G}H(t)^2. \tag{4.25}$$

If the energy density $\varepsilon(t)$ is greater than this value, the universe is positively curved ($\kappa = +1$). If $\varepsilon(t)$ is less than this value, the universe is negatively curved ($\kappa = -1$). Since we know the current value of the Hubble parameter to within 10%, we can compute the current value of the critical density to within 20%:

$$\varepsilon_{c,0} = \frac{3c^2}{8\pi G}H_0^2 = (8.3 \pm 1.7) \times 10^{-10}\,\text{J m}^{-3} = 5200 \pm 1000\,\text{MeV m}^{-3}. \tag{4.26}$$

[4]We also know from observations, as discussed earlier, that the radius of curvature must be comparable to or greater than the Hubble distance if the universe is *positively* curved.

The critical density is frequently written as the equivalent mass density,

$$\rho_{c,0} \equiv \frac{\varepsilon_{c,0}}{c^2} = (9.2 \pm 1.8) \times 10^{-27}\,\mathrm{kg\,m^{-3}} = (1.4 \pm 0.3) \times 10^{11}\,\mathrm{M_\odot\,Mpc^{-3}}. \tag{4.27}$$

Thus, the critical density is currently roughly equivalent to a density of one hydrogen atom per 200 liters, or 140 solar mass stars per cubic kiloparsec. This is definitely not a large density, by terrestrial standards. It's not even a large density by the standards of interstellar space within our galaxy, where even the hottest, most tenuous regions have a few protons per liter. However, keep in mind that most of the volume of the universe consists of intergalactic voids, where the density is extraordinarily low. When averaged over scales of 100 Mpc or more, the mean density of the universe, as it turns out, is close to the critical density.

In discussing the curvature of the universe, it is more convenient to use not the absolute density ε, but the ratio of the density to the critical density ε_c. Thus, when talking about the energy density of the universe, cosmologists often use the dimensionless *density parameter*

$$\Omega(t) \equiv \frac{\varepsilon(t)}{\varepsilon_c(t)}. \tag{4.28}$$

The most conservative limits on Ω—that is, limits that even the most belligerent cosmologist will hesitate to quarrel with—state that the current value of the density parameter lies in the range $0.1 < \Omega_0 < 2$.

In terms of the density parameter, the Friedmann equation can be written in yet another form:

$$1 - \Omega(t) = -\frac{\kappa c^2}{R_0^2 a(t)^2 H(t)^2}. \tag{4.29}$$

Note that, since the right-hand side of equation (4.29) cannot change sign as the universe expands, neither can the left-hand side. If $\Omega < 1$ at any time, it remains less than one for all time; similarly, if $\Omega > 1$ at any time, it remains greater than one for all time, and if $\Omega = 1$ at any time, $\Omega = 1$ at all times. A leopard can't change its spots; a universe governed by the Friedmann equation can't change the sign of its curvature. At the present moment, the relation among curvature, density, and expansion rate can be written in the form

$$1 - \Omega_0 = -\frac{\kappa c^2}{R_0^2 H_0^2}, \tag{4.30}$$

or

$$\frac{\kappa}{R_0^2} = \frac{H_0^2}{c^2}(\Omega_0 - 1). \tag{4.31}$$

If you know Ω_0, you know the sign of the curvature (κ). If, in addition, you know the Hubble distance, c/H_0, you can compute the radius of curvature (R_0).

4.2 ■ THE FLUID AND ACCELERATION EQUATIONS

Although the Friedmann equation is indeed important, it cannot, all by itself, tell us how the scale factor $a(t)$ evolves with time. Even if we had accurate boundary conditions (precise values for ε_0 and H_0, for instance), it still remains a single equation in two unknowns, $a(t)$ and $\varepsilon(t)$.[5]

We need another equation involving a and ε if we are to solve for a and ε as functions of time. The Friedmann equation, in the Newtonian approximation, is a statement of energy conservation; in particular, it says that the sum of the gravitational potential energy and the kinetic energy of expansion is constant. Energy conservation is a generally useful concept, so let's look at another manifestation of the same concept—the first law of thermodynamics—

$$dQ = dE + PdV, \tag{4.32}$$

where dQ is the heat flow into or out of a region, dE is the change in internal energy, P is the pressure, and dV is the change in volume of the region. This equation was applied in section 2.5 to a comoving volume filled with photons, but it applies equally well to a comoving volume filled with any sort of fluid. If the universe is perfectly homogeneous, then for any volume $dQ = 0$; that is, there is no bulk flow of heat. (Processes for which $dQ = 0$ are known as *adiabatic* processes. Saying that the expansion of the universe is adiabatic is also a statement about entropy. The change in entropy dS within a region is given by the relation $dS = dQ/T$; thus, an adiabatic process is one in which entropy is not increased. A homogeneous, isotropic expansion of the universe does not increase the universe's entropy.) Since $dQ = 0$ for a comoving volume as the universe expands, the first law of thermodynamics, as applied to the expanding universe, reduces to the form

$$\dot{E} + P\dot{V} = 0. \tag{4.33}$$

For concreteness, consider a sphere of comoving radius r_s expanding along with the universal expansion, so that its proper radius is $R_s(t) = a(t)r_s$. The volume of the sphere is

$$V(t) = \frac{4\pi}{3} r_s^3 a(t)^3, \tag{4.34}$$

so the rate of change of the sphere's volume is

$$\dot{V} = \frac{4\pi}{3} r_s^3 (3a^2\dot{a}) = V\left(3\frac{\dot{a}}{a}\right). \tag{4.35}$$

The internal energy of the sphere is

$$E(t) = V(t)\varepsilon(t), \tag{4.36}$$

[5]Or, if we prefer, we may take the unknown functions as $H(t)$ and $\Omega(t)$; in any case, there are two of them.

so the rate of change of the sphere's internal energy is

$$\dot{E} = V\dot{\varepsilon} + \dot{V}\varepsilon = V\left(\dot{\varepsilon} + 3\frac{\dot{a}}{a}\varepsilon\right). \tag{4.37}$$

Combining equations (4.33), (4.35), and (4.37), we find that the first law of thermodynamics in an expanding (or contracting) universe takes the form

$$V\left(\dot{\varepsilon} + 3\frac{\dot{a}}{a}\varepsilon + 3\frac{\dot{a}}{a}P\right) = 0, \tag{4.38}$$

or

$$\dot{\varepsilon} + 3\frac{\dot{a}}{a}(\varepsilon + P) = 0. \tag{4.39}$$

The above equation is called the *fluid equation*, and is the second of the key equations describing the expansion of the universe.[6]

The Friedmann equation and fluid equation are statements about energy conservation. By combining the two, we can derive an acceleration equation that tells how the expansion of the universe speeds up or slows down with time. The Friedmann equation (equation (4.13)), multiplied by a^2, takes the form

$$\dot{a}^2 = \frac{8\pi G}{3c^2}\varepsilon a^2 - \frac{\kappa c^2}{R_0^2}. \tag{4.40}$$

Taking the time derivative yields

$$2\dot{a}\ddot{a} = \frac{8\pi G}{3c^2}(\dot{\varepsilon}a^2 + 2\varepsilon a\dot{a}). \tag{4.41}$$

Dividing by $2\dot{a}a$ tells us

$$\frac{\ddot{a}}{a} = \frac{4\pi G}{3c^2}\left(\dot{\varepsilon}\frac{a}{\dot{a}} + 2\varepsilon\right). \tag{4.42}$$

Using the fluid equation (equation (4.39)), we may make the substitution

$$\dot{\varepsilon}\frac{a}{\dot{a}} = -3(\varepsilon + P) \tag{4.43}$$

to find the usual form of the *acceleration equation*,

$$\frac{\ddot{a}}{a} = -\frac{4\pi G}{3c^2}(\varepsilon + 3P). \tag{4.44}$$

Note that if the energy density ε is positive, then it provides a negative acceleration—that is, it decreases the value of $\dot{a}$ and reduces the relative velocity of any

[6]Write it on your forehead just underneath the Friedmann equation.

two points in the universe. The acceleration equation also includes the pressure P, associated with the material filling the universe.[7]

A gas made of ordinary baryonic matter has a positive pressure P, resulting from the random thermal motions of the molecules, atoms, or ions of which the gas is made. A gas of photons also has a positive pressure, as does a gas of neutrinos or WIMPs. The positive pressure associated with these components of the universe will cause the expansion to slow down. Suppose, though, that the universe had a component with a pressure

$$P < -\varepsilon/3. \tag{4.45}$$

Inspection of the acceleration equation (equation (4.44)) shows us that such a component will cause the expansion of the universe to speed up rather than slow down. A negative pressure (also called "tension") is certainly permissible by the laws of physics. Compress a piece of rubber, and its internal pressure will be positive; stretch the same piece of rubber, and its pressure will be negative. In cosmology, the much-discussed *cosmological constant* is a component of the universe with negative pressure. As we'll discuss in more detail in section 4.4, a cosmological constant has $P = -\varepsilon$, and thus causes a positive acceleration for the expansion of the universe.

4.3 ■ EQUATIONS OF STATE

To recap, we now have three key equations that describe how the universe expands. There's the Friedmann equation,

$$\left(\frac{\dot{a}}{a}\right)^2 = \frac{8\pi G}{3c^2}\varepsilon - \frac{\kappa c^2}{R_0^2 a^2}, \tag{4.46}$$

the fluid equation,

$$\dot{\varepsilon} + 3\frac{\dot{a}}{a}(\varepsilon + P) = 0, \tag{4.47}$$

and the acceleration equation,

$$\frac{\ddot{a}}{a} = -\frac{4\pi G}{3c^2}(\varepsilon + 3P). \tag{4.48}$$

Of these three equations, only two are independent, because equation (4.48), as we've just seen, can be derived from equations (4.46) and (4.47). Thus, we have a system of two independent equations in three unknowns—the functions $a(t)$, $\varepsilon(t)$, and $P(t)$. To solve for the scale factor, energy density, and pressure as a function of cosmic time, we need another equation. What we need is an *equation*

[7] Although we think of ε as an energy per unit volume and P as a force per unit area, they both have the same dimensionality: in SI units, $1\,\mathrm{J\,m^{-3}} = 1\,\mathrm{N\,m^{-2}} = 1\,\mathrm{kg\,m^{-1}\,s^{-2}}$.

of state; that is, a mathematical relation between the pressure and energy density of the stuff that fills up the universe. If only we had a relation of the form

$$P = P(\varepsilon), \tag{4.49}$$

life would be complete—or at least, our set of equations would be complete. We could then, given the appropriate boundary conditions, solve them to find how the universe expanded in the past, and how it will expand (or contract) in the future.

In general, equations of state can be dauntingly complicated. Condensed matter physicists frequently deal with substances in which the pressure is a complicated nonlinear function of the density. Fortunately, cosmology usually deals with dilute gases, for which the equation of state is simple. For substances of cosmological importance, the equation of state can be written in a simple linear form:

$$P = w\varepsilon, \tag{4.50}$$

where w is a dimensionless number.

Consider, for instance, a low-density gas of nonrelativistic massive particles. Nonrelativistic, in this case, means that the random thermal motions of the gas particles have peculiar velocities which are tiny compared to the speed of light. Such a nonrelativistic gas obeys the perfect gas law,

$$P = \frac{\rho}{\mu}kT, \tag{4.51}$$

where μ is the mean mass of the gas particles. The energy density ε of a nonrelativistic gas is almost entirely contributed by the mass of the gas particles: $\varepsilon \approx \rho c^2$. Thus, in terms of ε, the perfect gas law is

$$P \approx \frac{kT}{\mu c^2}\varepsilon. \tag{4.52}$$

For a nonrelativistic gas, the temperature T and the root mean square thermal velocity $\langle v^2 \rangle$ are associated by the relation

$$3kT = \mu\langle v^2 \rangle. \tag{4.53}$$

Thus, the equation of state for a nonrelativistic gas can be written in the form

$$P_{\rm nonrel} = w\varepsilon_{\rm nonrel}, \tag{4.54}$$

where

$$w \approx \frac{\langle v^2 \rangle}{3c^2} \ll 1. \tag{4.55}$$

Most of the gases we encounter in everyday life are nonrelativistic. For instance, in air at room temperature, nitrogen molecules are slow-poking along with a root mean square velocity of $\sim 500\,\mathrm{m\,s^{-1}}$, yielding $w \sim 10^{-12}$. Even in astronomical contexts, gases are mainly nonrelativistic at the present moment. Within a gas of ionized hydrogen, for instance, the electrons are nonrelativistic as long as $T \ll 6 \times 10^9$ K; the protons are nonrelativistic when $T \ll 10^{13}$ K.

A gas of photons, or other massless particles, is guaranteed to be relativistic. Although photons have no mass, they have momentum, and hence exert pressure. The equation of state of photons, or of any other relativistic gas, is

$$P_{\rm rel} = \frac{1}{3}\varepsilon_{\rm rel}. \tag{4.56}$$

(This relation has already been used in section 2.5, to compute how the Cosmic Microwave Background cools as the universe expands.) A gas of highly relativistic massive particles (with $\langle v^2 \rangle \sim c^2$) will also have $w = \frac{1}{3}$; a gas of mildly relativistic particles (with $0 < \langle v^2 \rangle < c^2$) will have $0 < w < \frac{1}{3}$.

The equation-of-state parameter w can't take on arbitrary values. Small perturbations in a substance with pressure P will travel at the speed of sound. For adiabatic perturbations in a gas with pressure P and energy density ε, the sound speed is given by the relation

$$c_s^2 = c^2 \left(\frac{dP}{d\varepsilon}\right). \tag{4.57}$$

In a substance with $w > 0$, the sound speed is thus $c_s = \sqrt{w}c$.[8] Sound waves cannot travel faster than the speed of light; if they did, you would be able to send a sound signal into the past, and violate causality. Thus, w is restricted to values $w \leq 1$.

Some values of w are of particular interest. For instance, the case $w = 0$ is of interest, because we know that our universe contains nonrelativistic matter. The case $w = \frac{1}{3}$ is of interest, because we know that our universe contains photons. For simplicity, we will refer to the component of the universe that consists of nonrelativistic particles (and hence has $w \approx 0$) as "matter," and the component that consists of photons and other relativistic particles (and hence has $w = \frac{1}{3}$) as "radiation." The case $w < -\frac{1}{3}$ is of interest, because a component with $w < -\frac{1}{3}$ will provide a positive acceleration ($\ddot{a} > 0$ in equation (4.48)). A component of the universe with $w < -\frac{1}{3}$ is sometimes referred to generically as "dark energy" (a phrase coined by the cosmologist Michael Turner). One form of dark energy is of special interest; some observational evidence, which we'll review in future chapters, indicates that our universe may contain a *cosmological constant.* A cosmological constant may be defined simply as a component of the universe that has $w = -1$, and hence has $P = -\varepsilon$. The cosmological constant, also designated by the Greek letter Λ, has had a controversial history, and is still the subject of debate. To learn why cosmologists have had such a long-standing love/hate affair with the cosmological constant Λ, it is necessary to make a brief historical review.

[8] In a substance with $w < 0$, the sound speed is an imaginary number; this implies that small pressure perturbations to a substance with negative pressure will not constitute stably propagating sound waves, but will have amplitudes that grow or decay with time.

4.4 ■ LEARNING TO LOVE LAMBDA

The cosmological constant Λ was first introduced by Albert Einstein. After publishing his first paper on general relativity in 1915, Einstein, naturally enough, wanted to apply his field equation to the real universe. He looked around, and noted that the universe contains both radiation and matter. Since Einstein, along with every other earthling of his time, was unaware of the existence of the Cosmic Microwave Background, he thought that most of the radiation in the universe was in the form of starlight. He also noted, quite correctly, that the energy density of starlight in our galaxy is much less than the rest energy density of the stars. Thus, Einstein concluded that the primary contribution to the energy density of the universe was from nonrelativistic matter, and that he could safely make the approximation that we live in a pressureless universe.

So far, Einstein was on the right track. However, in 1915, astronomers were unaware of the existence of the expansion of the universe. In fact, it was by no means settled that galaxies besides our own actually existed. After all, the sky is full of faint fuzzy patches of light. It took some time to sort out that some of the faint fuzzy patches are glowing clouds of gas within our galaxy and that some of them are galaxies in their own right, far beyond our own galaxy. Thus, when Einstein asked, "Is the universe expanding or contracting?" he looked, not at the motions of galaxies, but at the motions of stars within our galaxy. Einstein noted that some stars are moving toward us and that others are moving away from us, with no evidence that the galaxy is expanding or contracting.

The incomplete evidence available to Einstein led him to the belief that the universe is static—neither expanding nor contracting—and that it has a positive energy density but negligible pressure. Einstein then had to ask the question, "Can a universe filled with nonrelativistic matter, and nothing else, be static?" The answer to this question is "No!" A universe containing nothing but matter must, in general, be either expanding or contracting. The reason why this is true can be illustrated in a Newtonian context. If the mass density of the universe is ρ, then the gravitational potential Φ is given by Poisson's equation:

$$\nabla^2 \Phi = 4\pi G \rho. \tag{4.58}$$

The gravitational acceleration $\vec{a}$ at any point in space is then found by taking the gradient of the potential:

$$\vec{a} = -\vec{\nabla}\Phi. \tag{4.59}$$

In a static universe, $\vec{a}$ must vanish everywhere in space. Thus, the potential Φ must be constant in space. However, if Φ is constant, then (from equation (4.58))

$$\rho = \frac{1}{4\pi G}\nabla^2 \Phi = 0. \tag{4.60}$$

The only permissible static universe, in this analysis, is a totally empty universe. If you create a matter-filled universe that is initially static, then gravity will cause it to contract. If you create a matter-filled universe that is initially expanding, then it will either expand forever (if the Newtonian energy U is greater than or equal to zero) or reach a maximum radius and then collapse (if $U < 0$). Trying to make a matter-filled universe that doesn't expand or collapse is like throwing a ball into the air and expecting it to hover there.

How did Einstein surmount this problem? How did he reconcile the fact that the universe contains matter with his desire for a static universe? Basically, he added a fudge factor to the equations. In Newtonian terms, what he did was analogous to rewriting Poisson's equation in the form

$$\nabla^2 \Phi + \Lambda = 4\pi G\rho. \tag{4.61}$$

The new term, symbolized by the Greek letter Λ, came to be known as the *cosmological constant*. Note that it has dimensionality $(\text{time})^{-2}$. Introducing Λ into Poisson's equation allows the universe to be static if you set $\Lambda = 4\pi G\rho$.

In general relativistic terms, what Einstein did was to add an additional term, involving Λ, to his field equation (the relativistic equivalent of Poisson's equation). If the Friedmann equation is re-derived from Einstein's field equation, with the Λ term added, it becomes

$$\left(\frac{\dot{a}}{a}\right)^2 = \frac{8\pi G}{3c^2}\varepsilon - \frac{\kappa c^2}{R_0^2 a^2} + \frac{\Lambda}{3}. \tag{4.62}$$

The fluid equation is unaffected by the presence of a Λ term, so it still has the form

$$\dot{\varepsilon} + 3\frac{\dot{a}}{a}(\varepsilon + P) = 0. \tag{4.63}$$

With the Λ term present, the acceleration equation becomes

$$\frac{\ddot{a}}{a} = -\frac{4\pi G}{3c^2}(\varepsilon + 3P) + \frac{\Lambda}{3}. \tag{4.64}$$

A look at the Friedmann equation (4.62) tells us that adding the Λ term is equivalent to adding a new component to the universe with energy density

$$\varepsilon_\Lambda \equiv \frac{c^2}{8\pi G}\Lambda. \tag{4.65}$$

If Λ remains constant with time, then so does its associated energy density ε_Λ. The fluid equation (4.63) tells us that to have ε_Λ constant with time, the Λ term must have an associated pressure

$$P_\Lambda = -\varepsilon_\Lambda = -\frac{c^2}{8\pi G}\Lambda. \tag{4.66}$$

Thus, we can think of the cosmological constant as a component of the universe, which has a constant density ε_Λ and a constant pressure $P_\Lambda = -\varepsilon_\Lambda$.

By introducing a Λ term into his equations, Einstein got the static model universe he wanted. For the universe to remain static, both $\dot{a}$ and $\ddot{a}$ must be equal to zero. If $\ddot{a} = 0$, then in a universe with matter density ρ and cosmological constant Λ, the acceleration equation (4.64) reduces to

$$0 = -\frac{4\pi G}{3}\rho + \frac{\Lambda}{3}. \tag{4.67}$$

Thus, Einstein had to set $\Lambda = 4\pi G\rho$ in order to produce a static universe, just as in the Newtonian case. If $\dot{a} = 0$, the Friedmann equation (4.62) reduces to

$$0 = \frac{8\pi G}{3}\rho - \frac{\kappa c^2}{R_0^2} + \frac{\Lambda}{3} = 4\pi G\rho - \frac{\kappa c^2}{R_0^2}. \tag{4.68}$$

Einstein's static model therefore had to be positively curved ($\kappa = +1$), with a radius of curvature

$$R_0 = \frac{c}{2(\pi G\rho)^{1/2}} = \frac{c}{\Lambda^{1/2}}. \tag{4.69}$$

Although Einstein published the details of his static, positively curved, matter-filled model in the spring of 1917, he was dissatisfied with the model. He believed that the cosmological constant was "gravely detrimental to the formal beauty of the theory." In addition to its aesthetic shortcomings, the model had a practical defect; it was unstable. Although Einstein's static model was in equilibrium, with the repulsive force of Λ balancing the attractive force of ρ, it was an unstable equilibrium. Consider expanding Einstein's universe just a tiny bit. The energy density of Λ remains unchanged, but the energy density of matter drops. Thus, the repulsive force is greater than the attractive force, and the universe expands further. This causes the matter density to drop further, which causes the expansion to accelerate, which causes the matter density to drop further, and so forth. Expanding Einstein's static universe triggers runaway expansion; similarly, compressing it causes a runaway collapse.

Einstein was willing, even eager, to dispose of the "ugly" cosmological constant in his equations. Hubble's 1929 paper on the redshift–distance relation gave Einstein the necessary excuse for tossing Λ onto the rubbish heap. (Einstein later described the cosmological constant Λ as "the greatest blunder of my career.") Ironically, however, the same paper that caused Einstein to abandon the cosmological constant caused other scientists to embrace it. In his initial analysis, remember, Hubble badly underestimated the distance to galaxies, and hence overestimated the Hubble constant. Hubble's initial value of $H_0 = 500\,\text{km s}^{-1}\,\text{Mpc}^{-1}$ leads to a Hubble time of $H_0^{-1} = 2\,\text{Gyr}$, less than half the age of the Earth, as known from radioactive dating. How could cosmologists reconcile a short Hubble time with an old Earth? Some cosmologists pointed out that one way to increase the age of the universe for a given value of H_0^{-1} was to introduce a cosmological constant. If the value of Λ is large enough to make $\ddot{a} > 0$, then $\dot{a}$ was smaller in the past than it is now, and consequently the universe is older than H_0^{-1}.

If Λ has a value greater than $4\pi G\rho_0$, then the expansion of the universe is accelerating, and the universe can be arbitrarily old for a given value of H_0^{-1}. Since 1917, the cosmological constant has gone in and out of fashion, like sideburns or short skirts. It has been particularly fashionable during periods when the favored value of the Hubble time H_0^{-1} has been embarrassingly short compared to the estimated ages of astronomical objects. Currently, the cosmological constant is popular, thanks to observations, which we will discuss in Chapter 7, that indicate that the expansion of the universe does, indeed, have a positive acceleration.

A question that has been asked since the time of Einstein—and one which we've assiduously dodged until this moment—is "What is the physical cause of the cosmological constant?" In order to give Λ a real physical meaning, we need to identify some component of the universe whose energy density ε_Λ remains constant as the universe expands or contracts. Currently, the leading candidate for this component is the *vacuum energy*.

In classical physics, the idea of a vacuum having energy is nonsense. A vacuum, from the classical viewpoint, contains nothing; and as King Lear would say, "Nothing can come of nothing." In quantum physics, however, a vacuum is not a sterile void. The Heisenberg uncertainty principle permits particle–antiparticle pairs to spontaneously appear and then annihilate in an otherwise empty vacuum. The total energy ΔE and the lifetime Δt of these pairs of virtual particles must satisfy the relation[9]

$$\Delta E \Delta t \leq h. \tag{4.70}$$

Just as there's an energy density associated with the real particles in the universe, there is an energy density $\varepsilon_{\rm vac}$ associated with the virtual particle–antiparticle pairs. The vacuum density $\varepsilon_{\rm vac}$ is a quantum phenomenon that doesn't give a hoot about the expansion of the universe and is independent of time as the universe expands or contracts.

Unfortunately, computing the numerical value of $\varepsilon_{\rm vac}$ is an exercise in quantum field theory that has not yet been successfully completed. It has been suggested that the natural value for the vacuum energy density is the Planck energy density,

$$\varepsilon_{\rm vac} \sim \frac{E_P}{\ell_P^3} \qquad (???). \tag{4.71}$$

As we've seen in Chapter 1, the Planck energy is large by particle physics standards ($E_P = 1.2 \times 10^{28}\,{\rm eV}$), while the Planck length is small by anybody's standards ($\ell_P = 1.6 \times 10^{-35}\,{\rm m}$). This gives an energy density

$$\varepsilon_{\rm vac} \sim 3 \times 10^{133}\,{\rm eV\,m^{-3}} \qquad (!!!). \tag{4.72}$$

[9]The usual analogy that's made is to an embezzling bank teller who takes money from the till but who always replaces it before the auditor comes around. Naturally, the more money a teller is entrusted with, the more frequently the auditor checks up on her.

This is 124 orders of magnitude larger than the current critical density for our universe, and represents a spectacularly bad match between theory and observations. Obviously, we don't know much yet about the energy density of the vacuum! This is a situation where astronomers can help particle physicists, by deducing the value of ε_Λ from observations of the expansion of the universe. By looking at the universe at extremely large scales, we are indirectly examining the structure of the vacuum on extremely small scales.

SUGGESTED READING

Full references are given in the Annotated Bibliography on page 235.

Bernstein (1995), ch. 2: Friedmann equation, with applications

Harrison (2000), ch. 16: Newtonian derivation of Friedmann equation

Liddle (1999), ch. 3: Derivation of Friedmann, fluid, and acceleration equations

Rich (2001), ch. 4: "Relativistically correct" derivation of the Friedmann equation

PROBLEMS

4.1. Suppose the energy density of the cosmological constant is equal to the present critical density $\varepsilon_\Lambda = \varepsilon_{c,0} = 5200\,\text{MeV}\,\text{m}^{-3}$. What is the total energy of the cosmological constant within a sphere 1 AU in radius? What is the rest energy of the Sun ($E_\odot = M_\odot c^2$)? Comparing these two numbers, do you expect the cosmological constant to have a significant effect on the motion of planets within the solar system?

4.2. Consider Einstein's static universe, in which the attractive force of the matter density ρ is exactly balanced by the repulsive force of the cosmological constant, $\Lambda = 4\pi G\rho$. Suppose that some of the matter is converted into radiation (by stars, for instance). Will the universe start to expand or contract? Explain your answer.

4.3. If $\rho = 3 \times 10^{-27}\,\text{kg}\,\text{m}^{-3}$, what is the radius of curvature R_0 of Einstein's static universe? How long would it take a photon to circumnavigate such a universe?

4.4. Suppose that the universe were full of regulation baseballs, each of mass $m_{bb} = 0.145\,\text{kg}$ and radius $r_{bb} = 0.0369\,\text{m}$. If the baseballs were distributed uniformly throughout the universe, what number density of baseballs would be required to make the density equal to the critical density? (Assume nonrelativistic baseballs.) Given this density of baseballs, how far would you be able to see, on average, before your line of sight intersected a baseball? In fact, we can see galaxies at a distance $\sim c/H_0 \sim 4000\,\text{Mpc}$; does the transparency of the universe on this length scale place useful limits on the number density of intergalactic baseballs? (Note to readers outside North America or Japan: feel free to substitute regulation cricket balls, with $m_{cr} = 0.160\,\text{kg}$ and $r_{cr} = 0.0360\,\text{m}$.)

4.5. The principle of wave-particle duality tells us that a particle with momentum p has an associated de Broglie wavelength of $\lambda = h/p$; this wavelength increases as $\lambda \propto a$, as the universe expands. The total energy density of a gas of particles can be written as $\varepsilon = nE$, where n is the number density of particles, and E is the energy per particle. For simplicity, let's assume that all the gas particles have the same mass m and momentum p. The energy per particle is then simply

$$E = (m^2c^4 + p^2c^2)^{1/2} = (m^2c^4 + h^2c^2/\lambda^2)^{1/2} . \tag{4.73}$$

Compute the equation-of-state parameter w for this gas as a function of the scale factor a. Show that $w = \frac{1}{3}$ in the highly relativistic limit ($a \to 0$, $p \to \infty$) and that $w = 0$ in the highly nonrelativistic limit ($a \to \infty$, $p \to 0$).

CHAPTER 5

Single-Component Universes

In a spatially homogeneous and isotropic universe, the relation among the energy density $\varepsilon(t)$, the pressure $P(t)$, and the scale factor $a(t)$ is given by the Friedmann equation,

$$\left(\frac{\dot{a}}{a}\right)^2 = \frac{8\pi G}{3c^2}\varepsilon - \frac{\kappa c^2}{R_0^2 a^2}, \tag{5.1}$$

the fluid equation,

$$\dot{\varepsilon} + 3\frac{\dot{a}}{a}(\varepsilon + P) = 0, \tag{5.2}$$

and the equation of state,

$$P = w\varepsilon. \tag{5.3}$$

In principle, given the appropriate boundary conditions, we can solve equations (5.1), (5.2), and (5.3) to yield $\varepsilon(t)$, $P(t)$, and $a(t)$ for all times, past and future.

5.1 ■ EVOLUTION OF ENERGY DENSITY

In reality, the evolution of our universe is complicated by the fact that it contains different components with different equations of state. We know that the universe contains nonrelativistic matter and radiation—that's a conclusion as firm as the earth under your feet and as plain as daylight. Thus, the universe contains components with both $w = 0$ and $w = \frac{1}{3}$. It may well contain a cosmological constant, with $w = -1$. Moreover, the possibility exists that it may contain still more exotic components, with different values of w. Fortunately for the cause of simplicity, the energy density and pressure for the different components of the universe are additive. We may write the total energy density ε as the sum of the energy density of the different components:

$$\varepsilon = \sum_w \varepsilon_w, \tag{5.4}$$

where ε_w represents the energy density of the component with equation-of-state parameter w. The total pressure P is the sum of the pressures of the different components:

$$P = \sum_w P_w = \sum_w w\varepsilon_w. \tag{5.5}$$

Because the energy densities and pressures add in this way, the fluid equation (5.2) must hold for each component separately, as long as there is no interaction between the different components. If this is so, then the component with equation-of-state parameter w obeys the equation

$$\dot{\varepsilon}_w + 3\frac{\dot{a}}{a}(\varepsilon_w + P_w) = 0 \tag{5.6}$$

or

$$\dot{\varepsilon}_w + 3\frac{\dot{a}}{a}(1 + w)\varepsilon_w = 0. \tag{5.7}$$

Equation (5.7) can be rearranged to yield

$$\frac{d\varepsilon_w}{\varepsilon_w} = -3(1 + w)\frac{da}{a}. \tag{5.8}$$

If we assume that w is constant, then

$$\varepsilon_w(a) = \varepsilon_{w,0}a^{-3(1+w)}. \tag{5.9}$$

Here, we've used the usual normalization that $a_0 = 1$ at the present day, when the energy density of the w component is $\varepsilon_{w,0}$. Note that equation (5.9) is derived solely from the fluid equation and the equation of state; the Friedmann equation doesn't enter into it.

Starting from the general result of equation (5.9), we conclude that the energy density ε_m associated with nonrelativistic matter decreases as the universe expands with the dependence

$$\varepsilon_m(a) = \varepsilon_{m,0}/a^3. \tag{5.10}$$

The energy density in radiation, ε_r, drops at the steeper rate

$$\varepsilon_r(a) = \varepsilon_{r,0}/a^4. \tag{5.11}$$

Why this difference between matter and radiation? We may write the energy density of either component in the form $\varepsilon = nE$, where n is the number density of particles and E is the mean energy per particle. For both relativistic and nonrelativistic particles, the number density has the dependence $n \propto a^{-3}$ as the universe expands, assuming that particles are neither created nor destroyed.

Consider, once again, a sphere of comoving radius r_s, which expands along with the general expansion of the universe (Figure 5.1). When its proper radius

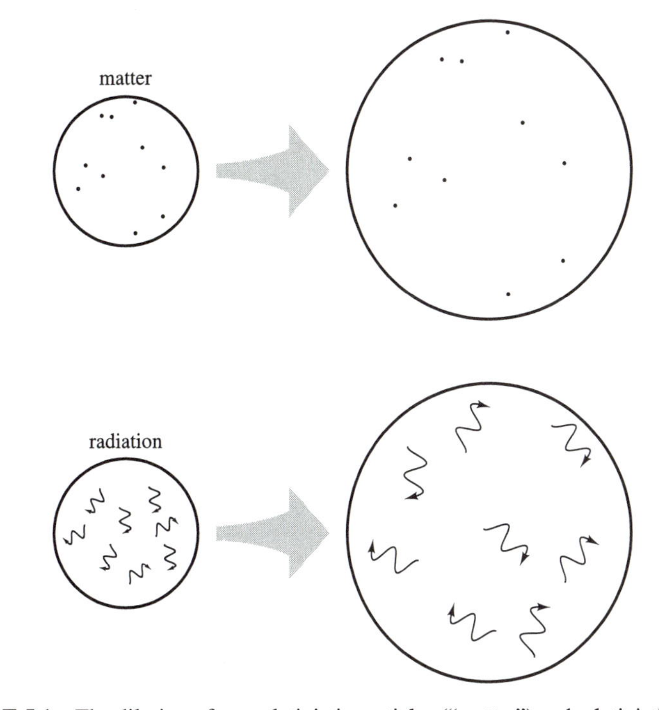

FIGURE 5.1 The dilution of nonrelativistic particles ("matter") and relativistic particles ("radiation") as the universe expands.

expands by a factor of 2, its volume increases by a factor of 8, and the density of particles that it contains falls to $\frac{1}{8}$ its previous value. The energy of nonrelativistic particles (shown in the upper panel of Figure 5.1) is contributed solely by their rest mass ($E = mc^2$) and remains constant as the universe expands. Thus, for nonrelativistic matter, $\varepsilon_m = nE = n(mc^2) \propto a^{-3}$. The energy of photons or other massless particles (shown in the lower panel of Figure 5.1) has the dependence $E = hc/\lambda \propto a^{-1}$, since, as shown in equation (3.45), $\lambda \propto a$ as the universe expands. Thus, for photons and other massless particles, $\varepsilon_r = nE = n(hc/\lambda) \propto a^{-3}a^{-1} \propto a^{-4}$.

Although we've explained why photons have an energy density $\varepsilon_r \propto a^{-4}$, the explanation required the assumption that photons are neither created nor destroyed. Such an assumption is not, strictly speaking, correct; photons are being created continuously. The Sun, for instance, is emitting roughly 10^{45} photons every second. However, so numerous are the photons of the Cosmic Microwave Background, it turns out that the energy density of the CMB is larger than the energy density of all the photons emitted by all the stars in the history of the universe. To see why this is true, remember, from section 2.5, that the present energy density of the CMB, which has a temperature $T_0 = 2.725\,\text{K}$, is

$$\varepsilon_{\text{CMB},0} = \alpha T_0^4 = 4.17 \times 10^{-14}\,\text{J}\,\text{m}^{-3} = 0.260\,\text{MeV}\,\text{m}^{-3} \tag{5.12}$$

yielding a density parameter for the CMB of

$$\Omega_{\text{CMB},0} \equiv \frac{\varepsilon_{\text{CMB},0}}{\varepsilon_{c,0}} = \frac{0.260\,\text{MeV}\,\text{m}^{-3}}{5200\,\text{MeV}\,\text{m}^{-3}} = 5.0 \times 10^{-5}. \tag{5.13}$$

Although the energy density of the CMB is small compared to the critical density, it is large compared to the energy density of starlight. Remember from section 2.3 that the present luminosity density of galaxies is

$$nL \approx 2 \times 10^8\,\text{L}_\odot\,\text{Mpc}^{-3} \approx 2.6 \times 10^{-33}\,\text{watts}\,\text{m}^{-3}. \tag{5.14}$$

As a *very* rough estimate, let's assume that galaxies have been emitting light at this rate for the entire age of the universe, $t_0 \approx H_0^{-1} \approx 14\,\text{Gyr} \approx 4.4 \times 10^{17}\,\text{s}$. This gives an energy density in starlight of

$$\begin{aligned} \varepsilon_{\text{starlight},0} \sim nLt_0 &\sim (2.6 \times 10^{-33}\,\text{J}\,\text{s}^{-1}\,\text{m}^{-3})(4.4 \times 10^{17}\,\text{s}) \\ &\sim 1 \times 10^{-15}\,\text{J}\,\text{m}^{-3} \\ &\sim 0.007\,\text{MeV}\,\text{m}^{-3}. \end{aligned} \tag{5.15}$$

Thus, the average energy density of starlight is currently only $\sim 3\%$ of the energy density of the CMB. In fact, the estimate given above is a very rough one indeed. Measurements of background radiation from ultraviolet wavelengths to the near infrared, which includes both direct starlight and starlight absorbed and reradiated by dust, yield the larger value $\varepsilon_{\text{starlight}}/\varepsilon_{\text{CMB}} \approx 0.1$. In the past, the ratio of starlight density to CMB density was even smaller than it is today. For most purposes, it is an acceptable approximation to ignore non-CMB photons when computing the mean energy density of photons in the universe.

The Cosmic Microwave Background, remember, is a relic of the time when the universe was hot and dense enough to be opaque to photons. If we extrapolate further back, we reach a time when the universe was hot and dense enough to be opaque to *neutrinos*. As a consequence, there should be a Cosmic Neutrino Background today, analogous to the Cosmic Microwave Background. The energy density in neutrinos should be comparable to, but not exactly equal to, the energy density in photons. A detailed calculation indicates that the energy density of each neutrino flavor should be

$$\varepsilon = \frac{7}{8}\left(\frac{4}{11}\right)^{4/3} \varepsilon_{\text{CMB}} \approx 0.227\,\varepsilon_{\text{CMB}}. \tag{5.16}$$

(The above result assumes that the neutrinos are relativistic, or, equivalently, that their energy is much greater than their rest energy $m_\nu c^2$.) The density parameter of the Cosmic Neutrino Background, taking into account all three flavors of neutrino, should then be

$$\Omega_\nu = 0.681\,\Omega_{\text{CMB}}, \tag{5.17}$$

as long as all neutrino flavors are relativistic. The mean energy per neutrino will be comparable to, but not exactly equal to, the mean energy per photon:

$$E_\nu \approx \frac{5 \times 10^{-4}\,\text{eV}}{a}, \tag{5.18}$$

as long as $E_\nu > m_\nu c^2$. When the mean energy of a particular neutrino species drops to $\sim m_\nu c^2$, then it makes the transition from being "radiation" to being "matter."

The neutrinos of the Cosmic Neutrino Background, I should note, have not yet been detected. Although neutrinos have been detected from the Sun and from Supernova 1987A, current technology permits the detection only of neutrinos with energy $E > 0.1\text{MeV}$, far more energetic than the neutrinos of the Cosmic Neutrino Background.

If all neutrino species are effectively massless today, with $m_\nu c^2 \ll 5 \times 10^{-4}\,\text{eV}$, then the present density parameter in radiation is

$$\Omega_{r,0} = \Omega_{\text{CMB},0} + \Omega_{\nu,0} = 5.0 \times 10^{-5} + 3.4 \times 10^{-5} = 8.4 \times 10^{-5}. \tag{5.19}$$

We know the energy density of the Cosmic Microwave Background with high precision. We can calculate theoretically what the energy density of the Cosmic Neutrino Background should be. Unfortunately, the total energy density of nonrelativistic matter is poorly known, as is the energy density of the cosmological constant. As we shall see in future chapters, the available evidence favors a universe in which the density parameter for matter is currently $\Omega_{m,0} \sim 0.3$, while the density parameter for the cosmological constant is currently $\Omega_{\Lambda,0} \sim 0.7$. Thus, when we want to employ a model that matches the observed properties of the real universe, we will use what I call the "Benchmark Model"; this model has $\Omega_{r,0} = 8.4 \times 10^{-5}$ in radiation, $\Omega_{m,0} = 0.3$ in nonrelativistic matter, and $\Omega_{\Lambda,0} = 1 - \Omega_{r,0} - \Omega_{m,0} \approx 0.7$ in a cosmological constant.[1]

In the Benchmark Model, at the present moment, the ratio of the energy density in Λ to the energy density in matter is

$$\frac{\varepsilon_{\Lambda,0}}{\varepsilon_{m,0}} = \frac{\Omega_{\Lambda,0}}{\Omega_{m,0}} \approx \frac{0.7}{0.3} \approx 2.3. \tag{5.20}$$

In the language of cosmologists, the cosmological constant is "dominant" over matter today in the Benchmark Model. In the past, however, when the scale factor was smaller, the ratio of densities was

$$\frac{\varepsilon_\Lambda(a)}{\varepsilon_m(a)} = \frac{\varepsilon_{\Lambda,0}}{\varepsilon_{m,0}/a^3} = \frac{\varepsilon_{\Lambda,0}}{\varepsilon_{m,0}}a^3. \tag{5.21}$$

If the universe has been expanding from an initial very dense state, at some moment in the past, the energy density of matter and Λ must have been equal. This

[1]Note that the Benchmark Model is spatially flat.

moment of matter-Λ equality occurred when the scale factor was

$$a_{m\Lambda} = \left(\frac{\Omega_{m,0}}{\Omega_{\Lambda,0}}\right)^{1/3} \approx \left(\frac{0.3}{0.7}\right)^{1/3} \approx 0.75. \tag{5.22}$$

Similarly, the ratio of the energy density in matter to the energy density in radiation is currently

$$\frac{\varepsilon_{m,0}}{\varepsilon_{r,0}} = \frac{\Omega_{m,0}}{\Omega_{r,0}} \approx \frac{0.3}{8.4 \times 10^{-5}} \approx 3600 \tag{5.23}$$

if all three neutrino flavors in the Cosmic Neutrino Background are still relativistic today; it's even larger if some or all of the neutrino flavors are currently nonrelativistic. Thus, matter is now strongly dominant over radiation. However, in the past, the ratio of matter density to energy density was

$$\frac{\varepsilon_m(a)}{\varepsilon_r(a)} = \frac{\varepsilon_{m,0}}{\varepsilon_{r,0}} a. \tag{5.24}$$

Thus, the moment of radiation-matter equality took place when the scale factor was

$$a_{rm} = \frac{\varepsilon_{r,0}}{\varepsilon_{m,0}} \approx \frac{1}{3600} \approx 2.8 \times 10^{-4}. \tag{5.25}$$

Note that as long as a neutrino's mass is $m_\nu c^2 \ll (3600)(5 \times 10^{-4}\,\text{eV}) \sim 2\,\text{eV}$, then it would have been relativistic at a scale factor $a = 1/3600$, and hence would have been "radiation" then even if it is "matter" today.

To generalize, if the universe contains different components with different values of w, equation (5.9) tells us that in the limit $a \to 0$, the component with the largest value of w is dominant. If the universe expands forever, then as $a \to \infty$, the component with the smallest value of w is dominant. The evidence indicates we live in a universe where radiation ($w = \frac{1}{3}$) was dominant during the early stages, followed by a period when matter ($w = 0$) was dominant. If the presently available evidence is correct, and we live in a universe described by the Benchmark Model, we have only recently entered a period when the cosmological constant Λ ($w = -1$) is dominant.

In a continuously expanding universe, the scale factor a is a monotonically increasing function of t. Thus, in a continuously expanding universe, the scale factor a can be used as a surrogate for the cosmic time t. We can refer, for instance, to the moment when $a = 2.8 \times 10^{-4}$ with the assurance that we are referring to a unique moment in the history of the universe. In addition, because of the simple relation between scale factor and redshift,

$$a = \frac{1}{1+z}, \tag{5.26}$$

cosmologists often use redshift as a surrogate for time. For example, they make statements such as, "Radiation-matter equality took place at a redshift $z_{rm} \approx$

3600." That is, radiation that was emitted at the time of radiation-matter equality is observed by us with its wavelength increased by a factor of 3600.

One reason why cosmologists use scale factor or redshift as a surrogate for time is that the conversion from a to t is not simple to calculate in a multiple-component universe like our own. In a universe with many components, the Friedmann equation can be written in the form

$$\dot{a}^2 = \frac{8\pi G}{3c^2} \sum_w \varepsilon_{w,0} a^{-1-3w} - \frac{\kappa c^2}{R_0^2}. \tag{5.27}$$

Each term on the right-hand side of equation (5.27) has a different dependence on scale factor; radiation contributes a term $\propto a^{-2}$, matter contributes a term $\propto a^{-1}$, curvature contributes a term independent of a, and the cosmological constant Λ contributes a term $\propto a^2$. Solving equation (5.27) for a multiple-component model, such as the Benchmark Model, does not yield a simple analytic form for $a(t)$. However, looking at simplified single-component universes, in which there is only one term on the right-hand side of equation (5.27), yields useful insight into the physics of an expanding universe.

5.2 ■ CURVATURE ONLY

A particularly simple universe is one that is empty—no radiation, no matter, no cosmological constant, no contribution to ε of any sort. For this universe, the Friedmann equation (5.27) takes the form

$$\dot{a}^2 = -\frac{\kappa c^2}{R_0^2}. \tag{5.28}$$

One solution to this equation has $\dot{a} = 0$ and $\kappa = 0$. An empty, static, spatially flat universe is a permissible solution to the Friedmann equation. This is the universe whose geometry is described by the Minkowski metric of equation (3.20), and in which all the transformations of special relativity hold true.

However, equation (5.28) tells us that it is also possible to have an empty universe with $\kappa = -1$. (Positively curved empty universes are forbidden, since that would require an imaginary value of $\dot{a}$ in equation (5.28).) A negatively curved empty universe must be expanding or contracting, with

$$\dot{a} = \pm \frac{c}{R_0}. \tag{5.29}$$

In an expanding empty universe, integration of this relation yields a scale factor of the form[2]

$$a(t) = \frac{t}{t_0}, \tag{5.30}$$

[2]Such an empty, negatively curved, expanding universe is sometimes called a *Milne universe*, after the first cosmologist to study its properties.

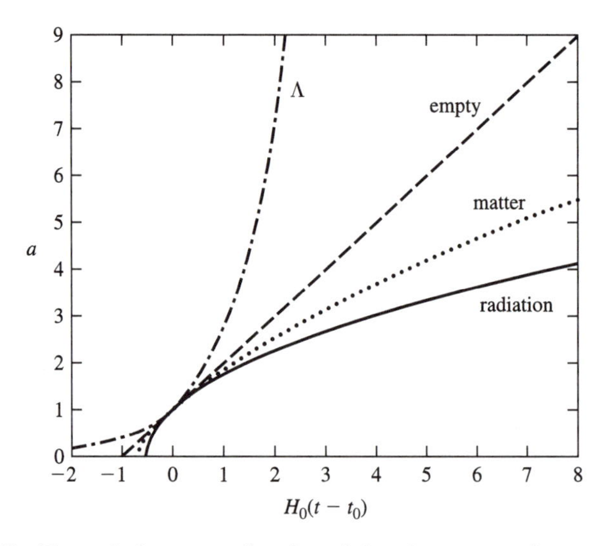

FIGURE 5.2 The scale factor as a function of time for an expanding, empty universe (dashed), a flat, matter-dominated universe (dotted), a flat, radiation-dominated universe (solid), and a flat, Λ-dominated universe (dot-dash).

where $t_0 = R_0/c$. In Newtonian terms, if there's no gravitational force at work, then the relative velocity of any two points is constant, and thus the scale factor a simply increases linearly with time in an empty universe. The scale factor in an empty, expanding universe is shown as the dashed line in Figure 5.2. Note that in an empty universe, $t_0 = H_0^{-1}$; with nothing to speed or slow the expansion, the age of the universe is exactly equal to the Hubble time.

An empty, expanding universe might seem nothing more than a mathematical curiosity.[3] However, if a universe has a density ε that is very small compared to the critical density ε_c (that is, if $\Omega \ll 1$), then the linear scale factor of equation (5.30) is a good approximation to the true scale factor.

Suppose you were in an expanding universe with a negligibly small value for the density parameter Ω, so that you could reasonably approximate it as an empty, negatively curved universe, with $t_0 = H_0^{-1} = R_0/c$. You observe a distant light source, such as a galaxy, which has a redshift z. The light you observe now, at $t = t_0$, was emitted at an earlier time, $t = t_e$. In an empty expanding universe,

$$1 + z = \frac{1}{a(t_e)} = \frac{t_0}{t_e}, \tag{5.31}$$

so it is easy to compute the time when the light you observe from the source was emitted:

$$t_e = \frac{t_0}{1+z} = \frac{H_0^{-1}}{1+z}. \tag{5.32}$$

[3]If a universe contains nothing, there will be no observers in it to detect the expansion.

When observing a galaxy with a redshift z, in addition to asking, "When was the light from that galaxy emitted?" you may also ask, "How far away is that galaxy?" In section 3.3 we saw that in any universe described by a Robertson–Walker metric, the current proper distance from an observer at the origin to a galaxy at coordinate location (r, θ, ϕ) is (see equation (3.28))

$$d_p(t_0) = a(t_0) \int_0^r dr = r. \tag{5.33}$$

Moreover, if light is emitted by the galaxy at time t_e and detected by the observer at time t_0, the null geodesic followed by the light satisfies equation (3.39):

$$c \int_{t_e}^{t_0} \frac{dt}{a(t)} = \int_0^r dr = r. \tag{5.34}$$

Thus, the current proper distance from you (the observer) to the galaxy (the light source) is

$$d_p(t_0) = c \int_{t_e}^{t_0} \frac{dt}{a(t)}. \tag{5.35}$$

Equation (5.35) holds true in any universe whose geometry is described by a Robertson–Walker metric. In the specific case of an empty expanding universe, $a(t) = t/t_0$, and thus

$$d_p(t_0) = ct_0 \int_{t_e}^{t_0} \frac{dt}{t} = ct_0 \ln\left(\frac{t_0}{t_e}\right). \tag{5.36}$$

Expressed in terms of the redshift z of the observed galaxy,

$$d_p(t_0) = \frac{c}{H_0} \ln(1 + z). \tag{5.37}$$

This relation is plotted as the dashed line in the upper panel of Figure 5.3. In the limit $z \ll 1$, there is a linear relation between d_p and z, as seen observationally in Hubble's law. In the limit $z \gg 1$, however, $d_p \propto \ln z$ in an empty expanding universe.

In an empty expanding universe, we can see objects that are currently at an arbitrarily large distance. However, at distances $d_p(t_0) \gg c/H_0$, the redshift increases exponentially with distance. At first glance, it may seem counterintuitive that we can see a light source at a proper distance much greater than c/H_0 when the age of the universe is only $1/H_0$. However, remember that $d_p(t_0)$ is the proper distance to the light source at the time of observation; at the time of *emission*, the proper distance $d_p(t_e)$ was smaller by a factor $a(t_e)/a(t_0) = 1/(1 + z)$. In an empty expanding universe, the proper distance at the time of emission was

$$d_p(t_e) = \frac{c}{H_0} \frac{\ln(1 + z)}{1 + z}, \tag{5.38}$$

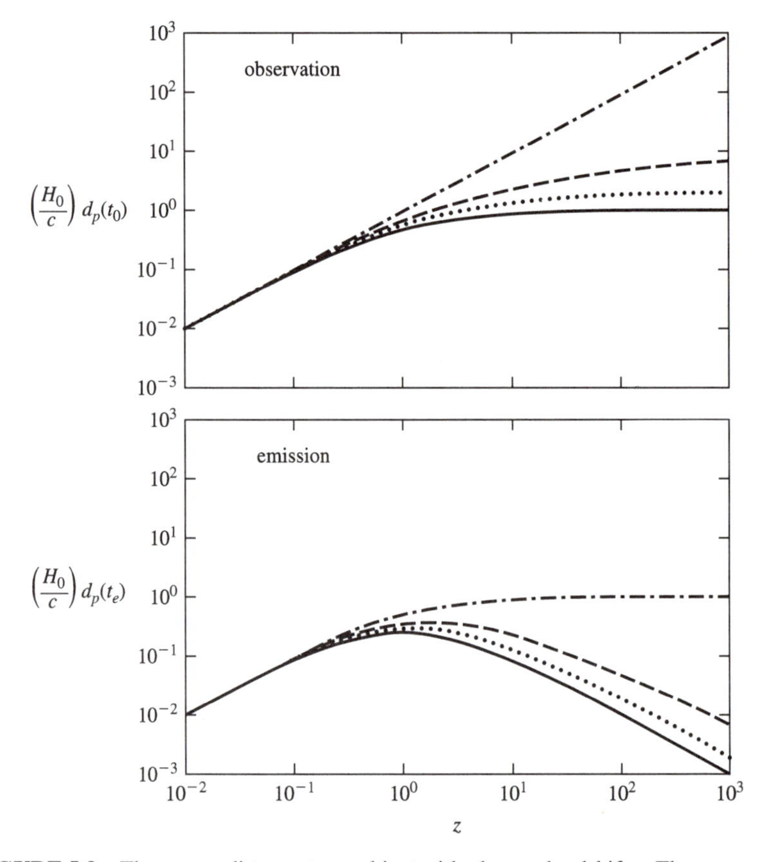

FIGURE 5.3 The proper distance to an object with observed redshift z. The upper panel shows the proper distance at the time the light is observed; the lower panel shows the proper distance at the time the light was emitted. The line types are the same as those of Figure 5.2.

shown as the dashed line in the lower panel of Figure 5.3. In an empty expanding universe, $d_p(t_e)$ has a maximum for objects with a redshift $z = e - 1 \approx 1.7$, where $d_p(t_e) = (1/e)(c/H_0) \approx 0.37(c/H_0)$. Objects with much higher redshifts are seen as they were very early in the history of the universe, when their proper distance from the observer was very small.

5.3 ■ SPATIALLY FLAT UNIVERSES

Setting the energy density ε equal to zero is one way of simplifying the Friedmann equation. Another way is to set $\kappa = 0$ and to demand that the universe contain only a single component, with a single value of w. In such a spatially flat, single-component universe, the Friedmann equation takes the simple form

$$\dot{a}^2 = \frac{8\pi G\varepsilon_0}{3c^2} a^{-(1+3w)}. \tag{5.39}$$

To solve this equation, we first make the educated guess that the scale factor has the power law form $a \propto t^q$. The left-hand side of equation (5.39) is then $\propto t^{2q-2}$, and the right-hand side is $\propto t^{-(1+3w)q}$, yielding the solution

$$q = \frac{2}{3+3w}, \tag{5.40}$$

with the restriction $w \neq -1$. With the proper normalization, the scale factor in a spatially flat, single-component universe is

$$a(t) = \left(\frac{t}{t_0}\right)^{2/(3+3w)}, \tag{5.41}$$

where the age of the universe, t_0, is linked to the present energy density by the relation

$$t_0 = \frac{1}{1+w}\left(\frac{c^2}{6\pi G\varepsilon_0}\right)^{1/2}. \tag{5.42}$$

The Hubble constant in such a universe is

$$H_0 \equiv \left(\frac{\dot{a}}{a}\right)_{t=t_0} = \frac{2}{3(1+w)} t_0^{-1}. \tag{5.43}$$

The age of the universe, in terms of the Hubble time, is then

$$t_0 = \frac{2}{3(1+w)} H_0^{-1}. \tag{5.44}$$

In a spatially flat universe, if $w > -\frac{1}{3}$, the universe is *younger* than the Hubble time. If $w < -\frac{1}{3}$, the universe is *older* than the Hubble time.

As a function of scale factor, the energy density of a component with equation-of-state parameter w is

$$\varepsilon(a) = \varepsilon_0 a^{-3(1+w)}, \tag{5.45}$$

so in a spatially flat universe with only a single component, the energy density as a function of time is (combining equations (5.41) and (5.45))

$$\varepsilon(t) = \varepsilon_0 \left(\frac{t}{t_0}\right)^{-2}, \tag{5.46}$$

regardless of the value of w. Making the substitutions

$$\varepsilon_0 = \varepsilon_{c,0} = \frac{3c^2}{8\pi G} H_0^2 \tag{5.47}$$

and

$$t_0 = \frac{2}{3(1+w)} H_0^{-1}, \tag{5.48}$$

equation (5.46) can be written in the form

$$\varepsilon(t) = \frac{1}{6\pi(1+w)^2} \frac{c^2}{G} t^{-2}. \tag{5.49}$$

Expressed in terms of Planck units (introduced in Chapter 1), this relation between energy density and cosmic time is

$$\varepsilon(t) = \frac{1}{6\pi(1+w)^2} \frac{E_P}{\ell_P^3} \left(\frac{t}{t_P}\right)^{-2}. \tag{5.50}$$

Suppose yourself to be in a spatially flat, single-component universe. If you see a galaxy, or other distant light source, with a redshift z, you can use the relation

$$1 + z = \frac{a(t_0)}{a(t_e)} = \left(\frac{t_0}{t_e}\right)^{2/(3+3w)} \tag{5.51}$$

to compute the time t_e at which the light from the distant galaxy was emitted:

$$t_e = \frac{t_0}{(1+z)^{3(1+w)/2}} = \frac{2}{3(1+w)H_0} \frac{1}{(1+z)^{3(1+w)/2}}. \tag{5.52}$$

The current proper distance to the galaxy is

$$d_p(t_0) = c \int_{t_e}^{t_0} \frac{dt}{a(t)} = ct_0 \frac{3(1+w)}{1+3w} [1 - (t_e/t_0)^{(1+3w)/(3+3w)}], \tag{5.53}$$

when $w \neq -\frac{1}{3}$. In terms of H_0 and z rather than t_0 and t_e, the current proper distance is

$$d_p(t_0) = \frac{c}{H_0} \frac{2}{1+3w} [1 - (1+z)^{-(1+3w)/2}]. \tag{5.54}$$

The most distant object you can see (in theory) is one for which the light emitted at $t = 0$ is just now reaching us at $t = t_0$. The proper distance (at the time of observation) to such an object is called the *horizon distance*.[4] Here on Earth,

[4]More technically, this is what's called the *particle horizon distance*; we'll continue to call it the horizon distance, for short.

the horizon is a circle centered on you, beyond which you cannot see because of the Earth's curvature. In the universe, the horizon is a spherical surface centered on you, beyond which you cannot see because light from more distant objects has not had time to reach you. In a universe described by a Robertson–Walker metric, the current horizon distance is

$$d_{\rm hor}(t_0) = c\int_0^{t_0} \frac{dt}{a(t)}. \tag{5.55}$$

In a spatially flat universe, the horizon distance has a finite value if $w > -\frac{1}{3}$. In such a case, computing the value of $d_p(t_0)$ in the limit $t_e \to 0$ (or, equivalently, $z \to \infty$) yields

$$d_{\rm hor}(t_0) = ct_0 \frac{3(1+w)}{1+3w} = \frac{c}{H_0}\frac{2}{1+3w}. \tag{5.56}$$

In a flat universe dominated by matter ($w = 0$) or by radiation ($w = \frac{1}{3}$), an observer can see only a finite portion of the infinite volume of the universe. The portion of the universe lying within the horizon for a particular observer is referred to as the *visible universe* for that observer. The visible universe consists of all points in space that have had sufficient time to send information, in the form of photons or other relativistic particles, to the observer. In other words, the visible universe consists of all points that are *causally connected* to the observer; nothing that happens outside the visible universe can have an effect on the observer.

In a flat universe with $w \leq -\frac{1}{3}$, the horizon distance is infinite, and all of space is causally connected to an observer. In such a universe with $w \leq -\frac{1}{3}$, you could see every point in space—assuming the universe was transparent, of course. However, for extremely distant points, you would see extremely redshifted versions of what they looked like extremely early in the history of the universe.

5.4 ■ MATTER ONLY

Let's now look at specific examples of spatially flat universes, starting with a universe containing only nonrelativistic matter ($w = 0$).[5] The age of such a universe is

$$t_0 = \frac{2}{3H_0}, \tag{5.57}$$

and the horizon distance is

$$d_{\rm hor}(t_0) = 3ct_0 = 2c/H_0. \tag{5.58}$$

[5]Such a universe is sometimes called an *Einstein-de Sitter universe*.

The scale factor, as a function of time, is

$$a_m(t) = \left(\frac{t}{t_0}\right)^{2/3}, \tag{5.59}$$

illustrated as the dotted line in Figure 5.2. If you see a galaxy with redshift z in a flat, matter-only universe, the proper distance to that galaxy, at the time of observation, is

$$d_p(t_0) = c\int_{t_e}^{t_0} \frac{dt}{(t/t_0)^{2/3}} = 3ct_0\left[1 - \left(\frac{t_e}{t_0}\right)^{1/3}\right] = \frac{2c}{H_0}\left[1 - \frac{1}{\sqrt{1+z}}\right], \tag{5.60}$$

illustrated as the dotted line in the upper panel of Figure 5.3. The proper distance at the time the light was emitted was smaller by a factor $1/(1+z)$:

$$d_p(t_e) = \frac{2c}{H_0(1+z)}\left[1 - \frac{1}{\sqrt{1+z}}\right], \tag{5.61}$$

illustrated as the dotted line in the lower panel of Figure 5.3. In a flat, matter-only universe, $d_p(t_e)$ has a maximum for galaxies with a redshift $z = \frac{5}{4}$, where $d_p(t_e) = (8/27)c/H_0$.

5.5 ■ RADIATION ONLY

The case of a spatially flat universe containing only *radiation* is of particular interest, since early in the history of our own universe, the radiation ($w = \frac{1}{3}$) term dominated the right-hand side of the Friedmann equation (see equation (5.27)). Thus, at early times—long before the time of radiation-matter equality—the universe was well described by a spatially flat, radiation-only model. In an expanding, flat universe containing only radiation, the age of the universe is

$$t_0 = \frac{1}{2H_0}, \tag{5.62}$$

and the horizon distance at t_0 is

$$d_{\mathrm{hor}}(t_0) = 2ct_0 = \frac{c}{H_0}. \tag{5.63}$$

In the special case of a flat, radiation-only universe, the horizon distance is exactly equal to the Hubble distance, which is not generally the case. The scale factor of a flat, radiation-only universe is

$$a(t) = \left(\frac{t}{t_0}\right)^{1/2}, \tag{5.64}$$

illustrated as the solid line in Figure 5.2. If at a time t_0 you observe a distant light source with redshift z in a flat, radiation-only universe, the proper distance to the light source will be

$$d_p(t_0) = c\int_{t_e}^{t_0} \frac{dt}{(t/t_0)^{1/2}} = 2ct_0\left[1 - \left(\frac{t_e}{t_0}\right)^{1/2}\right] = \frac{c}{H_0}\frac{z}{1+z}, \quad (5.65)$$

illustrated as the solid line in the upper panel of Figure 5.3. The proper distance at the time the light was emitted was

$$d_p(t_e) = \frac{c}{H_0(1+z)}\left[1 - \frac{1}{1+z}\right] = \frac{c}{H_0}\frac{z}{(1+z)^2}, \quad (5.66)$$

illustrated as the solid line in the lower panel of Figure 5.3. In a flat, radiation-dominated universe, $d_p(t_e)$ has a maximum for light sources with a redshift $z = 1$, where $d_p(t_e) = (1/4)c/H_0$.

The energy density in a flat, radiation-only universe is

$$\varepsilon_r(t) = \varepsilon_0\left(\frac{t}{t_0}\right)^{-2} = \frac{3}{32}\frac{E_P}{\ell_P^3}\left(\frac{t}{t_P}\right)^{-2} \approx 0.094\frac{E_P}{\ell_P^3}\left(\frac{t}{t_P}\right)^{-2}. \quad (5.67)$$

Thus, in the early stages of our universe, when radiation was strongly dominant, the energy density, measured in units of the Planck density ($E_P/\ell_P^3 \sim 3 \times 10^{133}\,\text{eV m}^{-3}$), was comparable to one over the square of the cosmic time, measured in units of the Planck time ($t_P \sim 5 \times 10^{-44}$ s). Using the blackbody relation between energy density and temperature, given in equations (2.26) and (2.27), we may assign a temperature to a universe dominated by blackbody radiation:

$$T(t) = \left(\frac{45}{32\pi^2}\right)^{1/4} T_P\left(\frac{t}{t_P}\right)^{-1/2} \approx 0.61T_P\left(\frac{t}{t_P}\right)^{-1/2}. \quad (5.68)$$

Here T_P is the Planck temperature, $T_P = 1.4 \times 10^{32}$ K. The mean energy per photon in a radiation-dominated universe is then

$$E_{\text{mean}}(t) \approx 2.70kT(t) \approx 1.66E_P\left(\frac{t}{t_P}\right)^{-1/2}, \quad (5.69)$$

and the number density of photons is (combining equations (5.67) and (5.69))

$$n(t) = \frac{\varepsilon_r(t)}{E_{\text{mean}}(t)} \approx \frac{0.057}{\ell_P^3}\left(\frac{t}{t_P}\right)^{-3/2}. \quad (5.70)$$

Note that in a flat, radiation-only universe, as $t \to 0$, $\varepsilon_r \to \infty$ (equation (5.67)). Thus, at the instant $t = 0$, the energy density of our own universe (well approximated as a flat, radiation-only model in its early stages) was infinite, according to this analysis; this infinite energy density was provided by an infinite number density of photons (equation (5.70)), each of infinite energy (equation (5.69)). Should we take these infinities seriously? Not really, since the assumptions of general relativity, on which the Friedmann equation is based, break down at $t \approx t_P$. Thus, extrapolating the results of this chapter to times earlier than the Planck time is not physically justified.

Why can't general relativity be used at times earlier than the Planck time? General relativity is a classical theory; that is, it does not take into account the effects of quantum mechanics. In cosmological contexts, general relativity assumes that the energy content of the universe is smooth down to arbitrarily small scales, instead of being parceled into individual quanta. As long as a radiation-dominated universe has many, many quanta, or photons, within a horizon distance, then the approximation of a smooth, continuous energy density is justifiable, and we may safely use the results of general relativity. However, if there are only a few photons within the visible universe, then quantum mechanical effects *must* be taken into account, and the classical results of general relativity no longer apply. In a flat, radiation-only universe, the horizon distance grows linearly with time:

$$d_{\rm hor}(t) = 2ct = 2\ell_P \left(\frac{t}{t_P}\right), \tag{5.71}$$

so the volume of the visible universe at time t is

$$V_{\rm hor}(t) = \frac{4\pi}{3} d_{\rm hor}^3 \approx 34\ell_P^3 \left(\frac{t}{t_P}\right)^3. \tag{5.72}$$

Combining equations (5.72) and (5.70), we find that the number of photons inside the horizon at time t is

$$N(t) = V_{\rm hor}(t) n(t) \approx 1.9 \left(\frac{t}{t_P}\right)^{3/2}. \tag{5.73}$$

The quantization of the universe can no longer be ignored when $N(t) \approx 1$, equivalent to a time $t \approx 0.7 t_P$.

To accurately describe the universe at its very earliest stages, prior to the Planck time, a theory of quantum gravity is needed. Unfortunately, a complete theory of quantum gravity does not yet exist. Consequently, in this book, we will not deal with times earlier than the Planck time, $t \sim t_P \sim 5 \times 10^{-44}$ s, when the number density of photons was $n \sim \ell_P^{-3} \sim 2 \times 10^{104}\,{\rm m}^{-3}$, and the mean photon energy was $E_{\rm mean} \sim E_P \sim 1 \times 10^{28}$ eV.

5.6 ■ LAMBDA ONLY

As seen in section 5.3, a spatially flat, single-component universe with $w \neq -1$ has a power-law dependence of scale factor on time:

$$a \propto t^{2/(3+3w)}. \tag{5.74}$$

Now, for the sake of completeness, consider the case with $w = -1$; that is, a universe in which the energy density is contributed by a cosmological constant Λ.[6] For such a flat, lambda-dominated universe, the Friedmann equation takes the form

$$\dot{a}^2 = \frac{8\pi G\varepsilon_\Lambda}{3c^2}a^2, \tag{5.75}$$

where ε_Λ is constant with time. This equation can be rewritten in the form

$$\dot{a} = H_0 a, \tag{5.76}$$

where

$$H_0 = \left(\frac{8\pi G\varepsilon_\Lambda}{3c^2}\right)^{1/2}. \tag{5.77}$$

The solution to equation (5.76) in an expanding universe is

$$a(t) = e^{H_0(t-t_0)}. \tag{5.78}$$

This scale factor is shown as the dot-dashed line in Figure 5.2. A spatially flat universe with nothing but a cosmological constant is exponentially expanding; we've seen an exponentially expanding universe before, in section 2.3, under the label "Steady State universe." In a Steady State universe, the density ε of the universe remains constant because of the continuous creation of real particles. If the cosmological constant Λ is provided by the vacuum energy, then the density ε of a lambda-dominated universe remains constant because of the continuous creation and annihilation of virtual particle-antiparticle pairs.

A flat universe containing nothing but a cosmological constant is infinitely old, and has an infinite horizon distance $d_{\rm hor}$. If, in a flat, lambda-only universe, you see a light source with a redshift z, the proper distance to the light source, at the time you observe it, is

$$d_p(t_0) = c\int_{t_e}^{t_0} e^{H_0(t_0-t)}dt = \frac{c}{H_0}[e^{H_0(t_0-t_e)} - 1] = \frac{c}{H_0}z, \tag{5.79}$$

shown as the dot-dashed line in the upper panel of Figure 5.3. The proper distance at the time the light was emitted was

$$d_p(t_e) = \frac{c}{H_0}\frac{z}{1+z}, \tag{5.80}$$

shown as the dot-dashed line in the lower panel of Figure 5.3.

Note that an exponentially growing universe, such as the flat lambda-dominated model, is the only universe for which $d_p(t_0)$ is linearly proportional to z for all values of z. In other universes, the relation $d_p(t_0) \propto z$ holds true only in the limit

[6] Such a universe is sometimes called a *de Sitter universe*.

$z \ll 1$. Note also that in the limit $z \to \infty$, $d_p(t_0) \to \infty$ but $d_p(t_e) \to c/H_0$. In a flat, lambda-dominated universe, highly redshifted objects ($z \gg 1$) are at very large distances ($d_p(t_0) \gg c/H_0$) at the time of observation; the observer sees them as they were just before they reached a proper distance c/H_0. Once a light source is more than a Hubble distance from the observer, its recession velocity is greater than the speed of light, and photons from the light source can no longer reach the observer.

The simple models that we've examined in this chapter—empty universes, or flat universes with a single component—continue to expand forever if they are expanding at $t = t_0$. Is it possible to have universes that stop expanding, then start to collapse? Is it possible to have universes in which the scale factor is not a simple power-law or exponential function of time? The short answer to these questions is "yes." To study universes with more complicated behavior, however, it is necessary to put aside our simple toy universes, with a single term on the right-hand side of the Friedmann equation, and look at complicated toy universes, with multiple terms on the right-hand side of the Friedmann equation.

SUGGESTED READING

Full references are given in the Annotated Bibiography on page 235.

Liddle (1999), ch. 4: Flat universes, both matter-only and radiation-only

PROBLEMS

5.6. The predicted number of neutrinos in the Cosmic Neutrino Background is $n_\nu = (3/11)n_\gamma = 1.12 \times 10^8\,\mathrm{m}^{-3}$ for each of the three species of neutrino. About how many cosmic neutrinos are inside your body right now? What must be the sum of the neutrino masses, $m(\nu_e) + m(\nu_\mu) + m(\nu_\tau)$, in order for the density of the Cosmic Neutrino Background to be equal to the critical density, $\varepsilon_{c,0}$?

5.7. A light source in a flat, single-component universe has a redshift z when observed at a time t_0. Show that the observed redshift changes at a rate

$$\frac{dz}{dt_0} = H_0(1+z) - H_0(1+z)^{3(1+w)/2}. \tag{5.81}$$

For what values of w does the redshift decrease with time? For what values of w does the redshift increase with time?

5.8. Suppose you are in a flat, matter-only universe that has a Hubble constant $H_0 = 70\,\mathrm{km\,s^{-1}\,Mpc^{-1}}$. You observe a galaxy with $z = 1$. How long will you have to keep observing the galaxy to see its redshift change by one part in 10^6? [Hint: use the result from the previous problem.]

5.9. In a flat universe with $H_0 = 70\,\text{km}\,\text{s}^{-1}\,\text{Mpc}^{-1}$, you observe a galaxy at a redshift $z = 7$. What is the current proper distance to the galaxy, $d_p(t_0)$, if the universe contains only radiation? What is $d_p(t_0)$ if the universe contains only matter? What is $d_p(t_0)$ if the universe contains only a cosmological constant? What was the proper distance at the time the light was emitted, $d_p(t_e)$, if the universe contains only radiation? What was $d_p(t_e)$ if the universe contains only matter? What was $d_p(t_e)$ if the universe contains only a cosmological constant?

CHAPTER 6

Multiple-Component Universes

The Friedmann equation, in general, can be written in the form

$$H(t)^2 = \frac{8\pi G}{3c^2}\varepsilon(t) - \frac{\kappa c^2}{R_0^2 a(t)^2}, \tag{6.1}$$

where $H \equiv \dot{a}/a$, and $\varepsilon(t)$ is the energy density contributed by all the components of the universe, including the cosmological constant. Equation (4.31) tells us the relation among κ, R_0, H_0, and Ω_0,

$$\frac{\kappa}{R_0^2} = \frac{H_0^2}{c^2}(\Omega_0 - 1), \tag{6.2}$$

so we can rewrite the Friedmann equation without explicitly including the curvature:

$$H(t)^2 = \frac{8\pi G}{3c^2}\varepsilon(t) - \frac{H_0^2}{a(t)^2}(\Omega_0 - 1). \tag{6.3}$$

Dividing by H_0^2, this becomes

$$\frac{H(t)^2}{H_0^2} = \frac{\varepsilon(t)}{\varepsilon_{c,0}} + \frac{1 - \Omega_0}{a(t)^2}, \tag{6.4}$$

where the critical density today is

$$\varepsilon_{c,0} \equiv \frac{3c^2 H_0^2}{8\pi G}. \tag{6.5}$$

We know that our universe contains matter, for which the energy density ε_m has the dependence $\varepsilon_m = \varepsilon_{m,0}/a^3$, and radiation, for which the energy density has the dependence $\varepsilon_r = \varepsilon_{r,0}/a^4$. Current evidence seems to indicate the presence of a cosmological constant, with energy density $\varepsilon_\Lambda = \varepsilon_{\Lambda,0} =$ constant. It is certainly possible that the universe contains other components as well. For instance, as the 21st century began, some cosmologists were investigating the properties of "quintessence," a component of the universe whose equation-of-state parameter can lie in the range $-1 < w < -\frac{1}{3}$, giving a universe with $\ddot{a} > 0$. However, in the absence of strong evidence for the existence of "quintessence," we will only consider the contributions of matter ($w = 0$), radiation ($w = \frac{1}{3}$), and the cosmological constant Λ ($w = -1$).

In our universe, we expect the Friedmann equation (6.4) to take the form

$$\frac{H^2}{H_0^2} = \frac{\Omega_{r,0}}{a^4} + \frac{\Omega_{m,0}}{a^3} + \Omega_{\Lambda,0} + \frac{1-\Omega_0}{a^2}, \tag{6.6}$$

where $\Omega_{r,0} = \varepsilon_{r,0}/\varepsilon_{c,0}$, $\Omega_{m,0} = \varepsilon_{m,0}/\varepsilon_{c,0}$, $\Omega_{\Lambda,0} = \varepsilon_{\Lambda,0}/\varepsilon_{c,0}$, and $\Omega_0 = \Omega_{r,0} + \Omega_{m,0} + \Omega_{\Lambda,0}$. The Benchmark Model, introduced in the previous chapter as a model consistent with all available data, has $\Omega_0 = 1$, and hence is spatially flat. However, although a perfectly flat universe is consistent with the data, it is not *demanded* by the data. Thus, prudence dictates that we should keep in mind the possibility that the curvature term, $(1 - \Omega_0)/a^2$ in equation (6.6), might be nonzero.

Since $H = \dot{a}/a$, multiplying equation (6.6) by a^2, then taking the square root, yields

$$H_0^{-1}\dot{a} = \left[\frac{\Omega_{r,0}}{a^2} + \frac{\Omega_{m,0}}{a} + \Omega_{\Lambda,0}a^2 + (1-\Omega_0)\right]^{1/2}. \tag{6.7}$$

The cosmic time t as a function of scale factor a can then be found by performing the integral

$$\int_0^a \frac{da}{[\Omega_{r,0}/a^2 + \Omega_{m,0}/a + \Omega_{\Lambda,0}a^2 + (1-\Omega_0)]^{1/2}} = H_0 t. \tag{6.8}$$

This is not a user-friendly integral: in the general case, it doesn't have a simple analytic solution. However, for given values of $\Omega_{r,0}$, $\Omega_{m,0}$, and $\Omega_{\Lambda,0}$, it can be integrated numerically.

In many circumstances, the integral in equation (6.8) has a simple analytic approximation to its solution. For instance, as noted in the previous chapter, in a universe with radiation, matter, curvature, and Λ, the radiation term dominates the expansion during the early stages of expansion. In this limit, equation (6.8) simplifies to

$$H_0 t \approx \int_0^a \frac{a\,da}{\sqrt{\Omega_{r,0}}} \approx \frac{1}{2\sqrt{\Omega_{r,0}}}a^2, \tag{6.9}$$

or

$$a(t) \approx (2\sqrt{\Omega_{r,0}}H_0 t)^{1/2}. \tag{6.10}$$

In the limit $\Omega_{r,0} = 1$, this reduces to the solution already found for a flat, radiation-only universe. If the universe continues to expand forever, then in the limit $a \to \infty$, the cosmological constant term will dominate the expansion. For some values of $\Omega_{r,0}$, $\Omega_{m,0}$, and $\Omega_{\Lambda,0}$, there will be intermediate epochs when the matter or the curvature dominates the expansion. For instance, in the Benchmark Model, where radiation-matter equality takes place at a scale factor $a_{rm} \approx 2.8 \times 10^{-4}$ and matter-lambda equality takes place at $a_{m\Lambda} \approx 0.75$, a matter-only universe is a fair approximation to reality when $a_{rm} \ll a \ll a_{m\Lambda}$.

However, during some epochs of the universe's expansion, two of the components are of comparable density, and provide terms of roughly equal size in the Friedmann equation. During these epochs, a single-component model is a poor description of the universe, and a two-component model must be utilized. For instance, for scale factors near $a_{rm} \approx 2.8 \times 10^{-4}$, the Benchmark Model is well approximated by a flat universe containing only radiation and matter. Such a universe is examined in section 6.4. For scale factors near $a_{m\Lambda} \approx 0.75$, the Benchmark Model is well approximated by a flat universe containing only matter and a cosmological constant. Such a universe is examined in section 6.2.

First, however, we will examine a universe that is of great historical interest to cosmology; a universe containing both matter and curvature (either negative or positive). After Einstein dismissed the cosmological constant as a blunder, and before astronomers had any clear idea what the value of the density parameter Ω was, they considered the possibility that the universe was negatively curved or positively curved, with the bulk of the density being provided by nonrelativistic matter. During the mid-twentieth century, cosmologists concentrated much of their interest on the study of curved, matter-dominated universes. In addition to being of historical interest, curved, matter-dominated universes provide useful physical insight into the interplay among curvature, expansion, and density.

6.1 ■ MATTER + CURVATURE

Consider a universe containing nothing but pressureless matter, with $w = 0$. If such a universe is spatially flat, then it expands with time, as demonstrated in section 5.4, with a scale factor

$$a(t) = \left(\frac{t}{t_0}\right)^{2/3} . \tag{6.11}$$

Such a flat, matter-only universe expands outward forever. Such a fate is sometimes known as the "Big Chill," since the temperature of the universe decreases monotonically with time as the universe expands. At this point, it is nearly obligatory for a cosmology text to quote T. S. Eliot: "This is the way the world ends / Not with a bang but a whimper."[1]

In a *curved* universe containing nothing but matter, the ultimate fate of the cosmos is intimately linked to the density parameter Ω_0. The Friedmann equation

[1]Interestingly, this quote is from Eliot's poem *The Hollow Men*, written, for the most part, in 1924, the year when Friedmann published his second paper on the expansion of the universe. However, this coincidence seems to be just that—a coincidence. Eliot did not keep up to date on the technical literature of cosmology.

in a curved, matter-dominated universe (equation (6.6)) can be written in the form

$$\frac{H(t)^2}{H_0^2} = \frac{\Omega_0}{a^3} + \frac{1 - \Omega_0}{a^2}, \tag{6.12}$$

since $\Omega_{m,0} = \Omega_0$ in such a universe. Suppose you are in a universe that is currently expanding ($H_0 > 0$) and contains nothing but nonrelativistic matter. If you ask the question, "Will the universe ever cease to expand?" then equation (6.12) enables you to answer that question. For the universe to cease expanding, there must be some moment at which $H(t) = 0$. Since the first term on the right-hand side of equation (6.12) is always positive, $H(t) = 0$ requires the second term on the right-hand side to be negative. This means that a matter-dominated universe will cease to expand if $\Omega_0 > 1$, and hence $\kappa = +1$. At the time of maximum expansion, $H(t) = 0$ and thus

$$0 = \frac{\Omega_0}{a_{\text{max}}^3} + \frac{1 - \Omega_0}{a_{\text{max}}^2}. \tag{6.13}$$

The scale factor at the time of maximum expansion will therefore be

$$a_{\text{max}} = \frac{\Omega_0}{\Omega_0 - 1}, \tag{6.14}$$

where Ω_0, remember, is the density parameter as measured at a scale factor $a = 1$.

Note that in equation (6.12), the Hubble parameter enters only as H^2. Thus, the contraction phase, after the universe reaches maximum expansion, is just the time reversal of the expansion phase.[2] Eventually, the $\Omega_0 > 1$ universe will collapse down to $a = 0$ (an event sometimes called the "Big Crunch") after a finite time $t = t_{\text{crunch}}$. A matter-dominated universe with $\Omega_0 > 1$ not only has finite spatial extent, but also has a finite duration in time; just as it began in a hot, dense state, so it will end in a hot, dense state. When such a universe is in its contracting stage, an observer will see galaxies with a *blueshift* proportional to their distance. As the universe approaches the Big Crunch, the cosmic microwave background will become a cosmic infrared background, then a cosmic visible background, then a cosmic ultraviolet background, then a cosmic x-ray background, then finally a cosmic gamma-ray background.

A matter-dominated universe with $\Omega_0 > 1$ will expand to a maximum scale factor a_{max}, then collapse in a Big Crunch. What is the ultimate fate of a matter-dominated universe with $\Omega_0 < 1$ and $\kappa = -1$? In the Friedmann equation for such a universe (equation (6.12)), *both* terms on the right-hand side are positive. Thus

[2]The contraction is a perfect time reversal of the expansion only when the universe is perfectly homogeneous and the expansion is perfectly adiabatic, or entropy-conserving. In a real, lumpy universe, entropy is not conserved on small scales. Stars, for instance, generate entropy as they emit photons. During the contraction phase of an $\Omega_0 > 1$ universe, small-scale entropy-producing processes will NOT be reversed. Stars will not absorb the photons they previously emitted; people will not live backward from grave to cradle.

TABLE 6.1 Curved, matter-dominated universes

Density	Curvature	Ultimate fate
$\Omega_0 < 1$	$\kappa = -1$	Big Chill ($a \propto t$)
$\Omega_0 = 1$	$\kappa = 0$	Big Chill ($a \propto t^{2/3}$)
$\Omega_0 > 1$	$\kappa = +1$	Big Crunch

if such a universe is expanding at a time $t = t_0$, it will continue to expand forever. At early times, when the scale factor is small ($a \ll \Omega_0/[1-\Omega_0]$), the matter term of the Friedmann equation will dominate, and the scale factor will grow at the rate $a \propto t^{2/3}$. Ultimately, however, the density of matter will be diluted far below the critical density, and the universe will expand like the negatively curved empty universe, with $a \propto t$.

If a universe contains nothing but matter, its curvature, its density, and its ultimate fate are closely linked, as shown in Table 6.1. At this point, the obligatory quote is from Robert Frost: "Some say the world will end in fire / Some say in ice."[3] In a matter-dominated universe, if the density is greater than the critical density, the universe will end in a fiery Big Crunch; if the density is less than or equal to the critical density, the universe will end in an icy Big Chill.

In a curved universe containing only matter, the scale factor $a(t)$ can be computed explicitly. The Friedmann equation can be written in the form

$$\frac{\dot{a}^2}{H_0^2} = \frac{\Omega_0}{a} + (1 - \Omega_0), \tag{6.15}$$

so the age t of the universe at a given scale factor a is given by the integral

$$H_0 t = \int_0^a \frac{da}{[\Omega_0/a + (1 - \Omega_0)]^{1/2}}. \tag{6.16}$$

When $\Omega_0 \neq 1$, the solution to this integral is most compactly written in a parametric form. The solution when $\Omega_0 > 1$ is

$$a(\theta) = \frac{1}{2}\frac{\Omega_0}{\Omega_0 - 1}(1 - \cos\theta) \tag{6.17}$$

and

$$t(\theta) = \frac{1}{2H_0}\frac{\Omega_0}{(\Omega_0 - 1)^{3/2}}(\theta - \sin\theta), \tag{6.18}$$

where the parameter θ runs from 0 to 2π. Given this parametric form, it is easy to show that the time that elapses between the Big Bang at $\theta = 0$ and the Big

[3]This is from Frost's poem *Fire and Ice*, first published in Harper's Magazine in December 1920. Unlike T. S. Eliot, Frost was keenly interested in astronomy, and frequently wrote poems on astronomical themes.

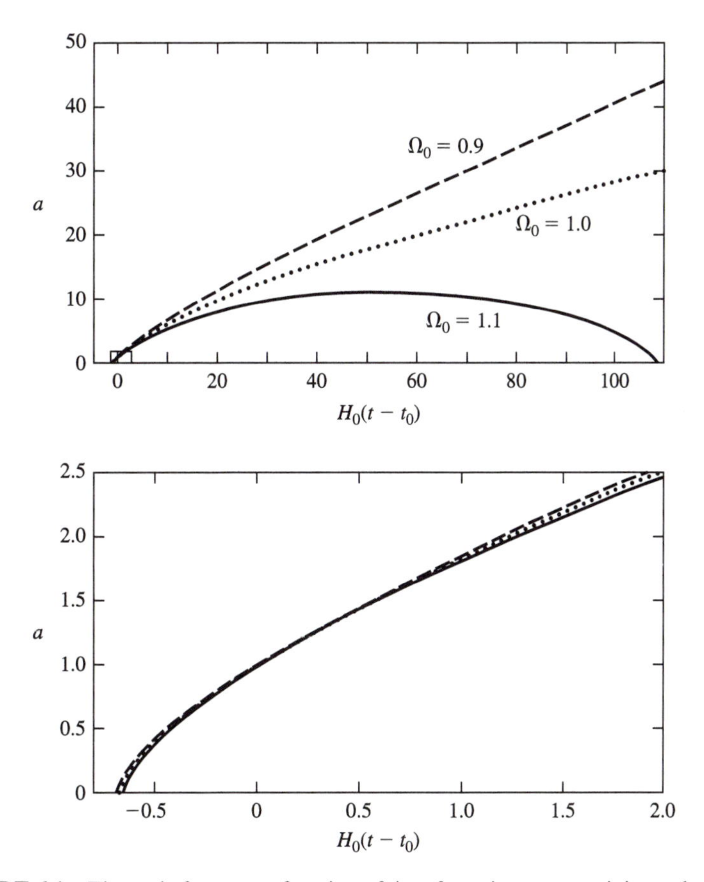

FIGURE 6.1 The scale factor as a function of time for universes containing only matter. The dotted line is $a(t)$ for a universe with $\Omega_0 = 1$ (flat); the dashed line is $a(t)$ for a universe with $\Omega_0 = 0.9$ (negatively curved); the solid line is $a(t)$ for a universe with $\Omega_0 = 1.1$ (positively curved). The bottom panel is a blow-up of the small rectangle near the lower left corner of the upper panel.

Crunch at $\theta = 2\pi$ is

$$t_{\text{crunch}} = \frac{\pi}{H_0} \frac{\Omega_0}{(\Omega_0 - 1)^{3/2}}. \tag{6.19}$$

A plot of a versus t in the case $\Omega_0 = 1.1$ is shown as the solid line in Figure 6.1. The $a \propto t^{2/3}$ behavior of an $\Omega_0 = 1$ universe is shown as the dotted line. The solution of equation (6.16) for the case $\Omega_0 < 1$ can be written in parametric form as

$$a(\eta) = \frac{1}{2} \frac{\Omega_0}{1 - \Omega_0} (\cosh \eta - 1) \tag{6.20}$$

and

$$t(\eta) = \frac{1}{2} \frac{\Omega_0}{(1 - \Omega_0)^{3/2}} (\sinh \eta - \eta), \tag{6.21}$$

where the parameter η runs from 0 to infinity. A plot of a versus t in the case $\Omega_0 = 0.9$ is shown as the dashed line in Figure 6.1. Note that although the ultimate fates of an $\Omega_0 = 0.9$ universe is very different from that of an $\Omega_0 = 1.1$ universe, as shown graphically in the upper panel of Figure 6.1, it is very difficult, at $t \sim t_0$, to tell a universe with Ω_0 slightly less than one from that with Ω_0 slightly greater than one. As shown in the lower panel of Figure 6.1, the scale factors of the $\Omega_0 = 1.1$ universe and the $\Omega_0 = 0.9$ universe start to diverge significantly only after a Hubble time or more.

Scientists sometimes joke that they are searching for a theory of the universe that is compact enough to fit on the front of a T-shirt. If the energy content of the universe is contributed almost entirely by nonrelativistic matter, then an appropriate T-shirt slogan would be:

DENSITY
IS
DESTINY!

If the density of matter is less than the critical value, then the destiny of the universe is an ever-expanding Big Chill; if the density is greater than the critical value, then the destiny is a recollapsing Big Crunch. Like all terse summaries of complex concepts, the slogan "Density is Destiny!" requires a qualifying footnote. In this case, the required footnote is "*if $\Lambda = 0$." If the universe has a cosmological constant (or more generally, any component with $w < -\frac{1}{3}$), then the equation Density = Destiny = Curvature no longer applies.

6.2 ■ MATTER + LAMBDA

Consider a universe that is spatially flat, but contains both matter and a cosmological constant.[4] If, at a given time $t = t_0$, the density parameter in matter is $\Omega_{m,0}$ and the density parameter in a cosmological constant Λ is $\Omega_{\Lambda,0}$, the requirement that space be flat tells us that

$$\Omega_{\Lambda,0} = 1 - \Omega_{m,0}, \tag{6.22}$$

and the Friedmann equation for the flat "matter plus lambda" universe reduces to

$$\frac{H^2}{H_0^2} = \frac{\Omega_{m,0}}{a^3} + (1 - \Omega_{m,0}). \tag{6.23}$$

[4]Such a universe is of particular interest to us, since it appears to be a close approximation to our own universe at the present day.

The first term on the right-hand side of equation (6.23) represents the contribution of matter, and is always positive. The second term represents the contribution of a cosmological constant; it is positive if $\Omega_{m,0} < 1$, implying $\Omega_{\Lambda,0} > 0$, and negative if $\Omega_{m,0} > 1$, implying $\Omega_{\Lambda,0} < 0$. Thus, a flat universe with $\Omega_{\Lambda,0} > 0$ will continue to expand forever if it is expanding at $t = t_0$; this is another example of a Big Chill universe. In a universe with $\Omega_{\Lambda,0} < 0$, however, the negative cosmological constant provides an *attractive* force, not the repulsive force of a positive cosmological constant. A flat universe with $\Omega_{\Lambda,0} < 0$ will cease to expand at a maximum scale factor

$$a_{\max} = \left(\frac{\Omega_{m,0}}{\Omega_{m,0} - 1} \right)^{1/3}, \tag{6.24}$$

and will collapse back down to $a = 0$ at a cosmic time

$$t_{\text{crunch}} = \frac{2\pi}{3H_0} \frac{1}{\sqrt{\Omega_{m,0} - 1}}. \tag{6.25}$$

For a given value of H_0, the larger the value of $\Omega_{m,0}$, the shorter the lifetime of the universe. For a flat, $\Omega_{\Lambda,0} < 0$ universe, the Friedmann equation (6.23) can be integrated to yield the analytic solution

$$H_0 t = \frac{2}{3\sqrt{\Omega_{m,0} - 1}} \sin^{-1} \left[\left(\frac{a}{a_{\max}} \right)^{3/2} \right]. \tag{6.26}$$

A plot of a versus t in the case $\Omega_{m,0} = 1.1$, $\Omega_{\Lambda,0} = -0.1$ is shown as the solid line in Figure 6.2. The $a \propto t^{2/3}$ behavior of a $\Omega_{m,0} = 1$, $\Omega_{\Lambda,0} = 0$ universe is

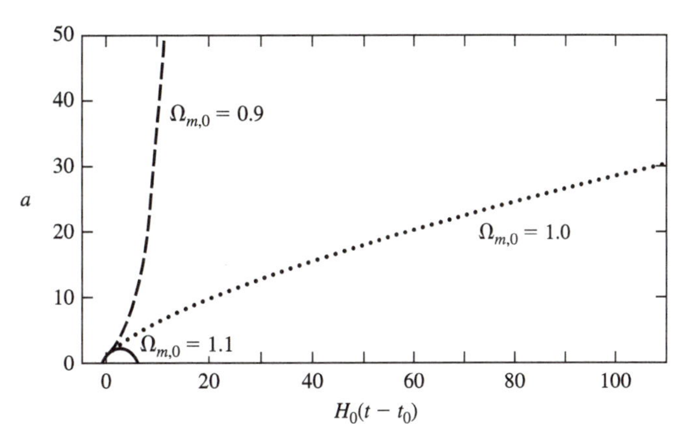

FIGURE 6.2 The scale factor as a function of time for flat universes containing both matter and a cosmological constant. The dotted line is $a(t)$ for a universe with $\Omega_{m,0} = 1$, $\Omega_{\Lambda,0} = 0$. The solid line is $a(t)$ for a universe with $\Omega_{m,0} = 1.1$, $\Omega_{\Lambda,0} = -0.1$. The dashed line is $a(t)$ for a universe with $\Omega_{m,0} = 0.9$, $\Omega_{\Lambda,0} = 0.1$.

shown, for comparison, as the dotted line. A flat universe with $\Omega_{\Lambda,0} < 0$ ends in a Big Crunch, reminiscent of that for a positively curved, matter-only universe. However, with a negative cosmological constant providing an attractive force, the lifetime of a flat universe with $\Omega_{\Lambda,0} < 0$ is exceptionally short. For instance, we have seen that a positively curved universe with $\Omega_{m,0} = 1.1$ undergoes a Big Crunch after a lifetime $t_{\text{crunch}} \approx 110 H_0^{-1}$ (see Figure 6.1). However, a flat universe with $\Omega_{m,0} = 1.1$ and $\Omega_{\Lambda,0} = -0.1$ has a lifetime $t_{\text{crunch}} \approx 7 H_0^{-1}$. As soon as the universe becomes lambda-dominated, the negative cosmological constant causes a rapid deceleration of the universe's expansion.

Although a negative cosmological constant is permitted by the laws of physics, it appears that we live in a universe where the cosmological constant is non-negative. In a flat universe with $\Omega_{m,0} < 1$ and $\Omega_{\Lambda,0} > 0$, the density contributions of matter and the cosmological constant are equal at the scale factor (equation (5.22)):

$$a_{m\Lambda} = \left(\frac{\Omega_{m,0}}{\Omega_{\Lambda,0}}\right)^{1/3} = \left(\frac{\Omega_{m,0}}{1-\Omega_{m,0}}\right)^{1/3}. \tag{6.27}$$

For a flat, $\Omega_{\Lambda,0} > 0$ universe, the Friedmann equation can be integrated to yield the analytic solution

$$H_0 t = \frac{2}{3\sqrt{1-\Omega_{m,0}}} \ln\left[\left(\frac{a}{a_{m\Lambda}}\right)^{3/2} + \sqrt{1+\left(\frac{a}{a_{m\Lambda}}\right)^3}\right]. \tag{6.28}$$

A plot of a versus t in the case $\Omega_{m,0} = 0.9$, $\Omega_{\Lambda,0} = 0.1$ is shown as the dashed line in Figure 6.2. At early times, when $a \ll a_{m\Lambda}$, equation (6.28) reduces to the relation

$$a(t) \approx \left(\frac{3}{2}\sqrt{\Omega_{m,0}}\, H_0 t\right)^{2/3}, \tag{6.29}$$

giving the $a \propto t^{2/3}$ dependence required for a flat, matter-dominated universe. At late times, when $a \gg a_{m\Lambda}$, equation (6.28) reduces to

$$a(t) \approx a_{m\Lambda} \exp(\sqrt{1-\Omega_{m,0}}\, H_0 t), \tag{6.30}$$

giving the $a \propto e^{Kt}$ dependence required for a flat, lambda-dominated universe. Suppose you are in a flat universe containing nothing but matter and a cosmological constant; if you measure H_0 and $\Omega_{m,0}$, then equation (6.28) tells you that the age of the universe is

$$t_0 = \frac{2H_0^{-1}}{3\sqrt{1-\Omega_{m,0}}} \ln\left[\frac{\sqrt{1-\Omega_{m,0}}+1}{\sqrt{\Omega_{m,0}}}\right]. \tag{6.31}$$

If we approximate our own universe as having $\Omega_{m,0} = 0.3$ and $\Omega_{\Lambda,0} = 0.7$ (ignoring the contribution of radiation) we find that its current age is

$$t_0 = 0.964 H_0^{-1} = 13.5 \pm 1.3\,\text{Gyr}, \tag{6.32}$$

assuming $H_0 = 70 \pm 7\,\mathrm{km\,s^{-1}\,Mpc^{-1}}$. (We'll see in section 6.5 that ignoring the radiation content of the universe has an insignificant effect on our estimate of t_0.) The age at which matter and the cosmological constant had equal energy density was

$$t_{m\Lambda} = \frac{2H_0^{-1}}{3\sqrt{1-\Omega_{m,0}}}\ln[1+\sqrt{2}] = 0.702H_0^{-1} = 9.8 \pm 1.0\,\mathrm{Gyr}. \tag{6.33}$$

Thus, if our universe is well fit by the Benchmark Model, with $\Omega_{m,0} = 0.3$ and $\Omega_{\Lambda,0} \approx 0.7$, then the cosmological constant has been the dominant component of the universe for the last four billion years or so.

6.3 ■ MATTER + CURVATURE + LAMBDA

If a universe contains both matter and a cosmological constant, then the formula "density = destiny = curvature" no longer holds. A flat universe with $\Omega_{m,0} > 1$ and $\Omega_{\Lambda,0} < 0$, as shown in the previous section, is infinite in spatial extent, but has a finite duration in time. By contrast, a flat universe with $\Omega_{m,0} \leq 1$ and $\Omega_{\Lambda,0} \geq 0$ extends to infinity both in space and in time. If a universe containing both matter and lambda is curved ($\kappa \neq 0$) rather than flat, then a wide range of behaviors is possible for the function $a(t)$. For instance, in section 4.4, we encountered Einstein's static model, in which $\kappa = +1$ and $\varepsilon_\Lambda = \varepsilon_m/2$. A universe described by Einstein's static model is finite in spatial extent, but has infinite duration in time.

By choosing different values of $\Omega_{m,0}$ and $\Omega_{\Lambda,0}$, without constraining the universe to be flat, we can create model universes with scale factors $a(t)$ that exhibit very interesting behavior. Start by writing down the Friedmann equation for a curved universe with both matter and a cosmological constant:

$$\frac{H^2}{H_0^2} = \frac{\Omega_{m,0}}{a^3} + \frac{1-\Omega_{m,0}-\Omega_{\Lambda,0}}{a^2} + \Omega_{\Lambda,0}. \tag{6.34}$$

If $\Omega_{m,0} > 0$ and $\Omega_{\Lambda,0} > 0$, then both the first and last term on the right-hand side of equation (6.34) are positive. However, if $\Omega_{m,0} + \Omega_{\Lambda,0} > 1$, so that the universe is positively curved, then the central term on the right-hand side is negative. As a result, for some choices of $\Omega_{m,0}$ and $\Omega_{\Lambda,0}$, the value of H^2 will be positive for small values of a (where matter dominates) and for large values of a (where Λ dominates), but will be negative for intermediate values of a (where the curvature term dominates). Since negative values of H^2 are unphysical, this means that these universes have a forbidden range of scale factors. Suppose such a universe starts out with $a \gg 1$ and $H < 0$; that is, it is contracting from a low-density, Λ-dominated state. As the universe contracts, however, the negative curvature term in equation (6.34) becomes dominant, causing the contraction to stop at a minimum scale factor $a = a_{\mathrm{min}}$, and then expand outward again in a "Big Bounce."

Thus, it is possible to have a universe that expands outward at late times, but never had an initial Big Bang, with $a = 0$ at $t = 0$. Another possibility, if the values of $\Omega_{m,0}$ and $\Omega_{\Lambda,0}$ are chosen just right, is a "loitering" universe.[5] Such a universe starts in a matter-dominated state, expanding outward with $a \propto t^{2/3}$. Then, however, it enters a stage (called the loitering stage) in which a is very nearly constant for a long period of time. During this time it is almost—but not quite—Einstein's static universe. After the loitering stage, the cosmological constant takes over, and the universe starts to expand exponentially.

Figure 6.3 shows the general behavior of the scale factor $a(t)$ as a function of $\Omega_{m,0}$ and $\Omega_{\Lambda,0}$. In the region labeled "Big Crunch," the universe starts with $a = 0$ at $t = 0$, reaches a maximum scale factor $a_{\max}$, then recollapses to $a = 0$ at a finite time $t = t_{\text{crunch}}$. Note that Big Crunch universes can be positively curved,

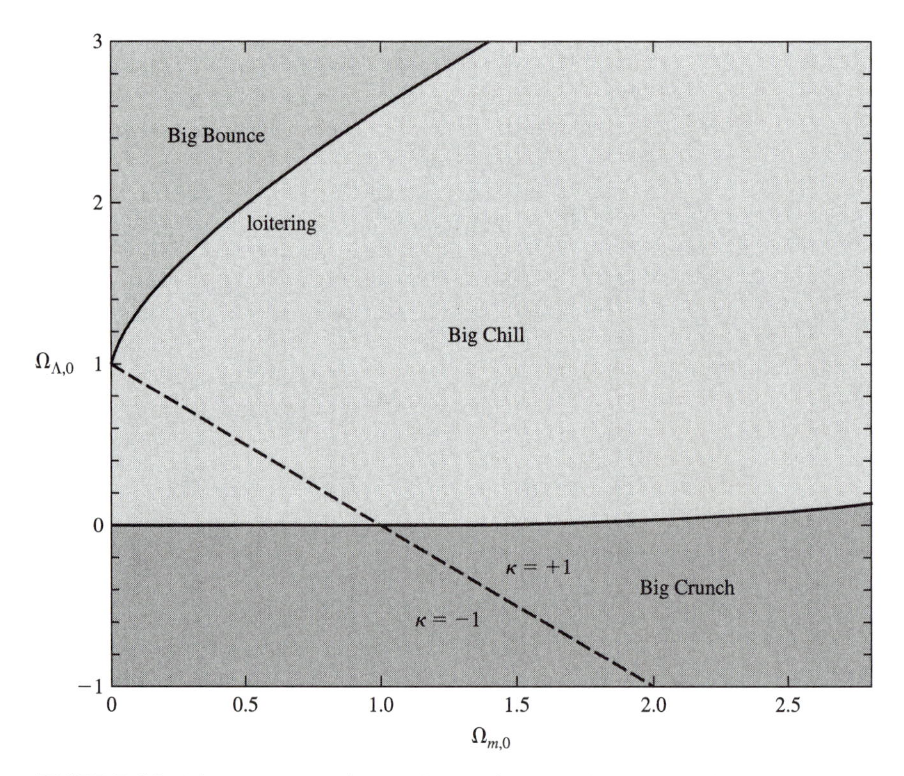

FIGURE 6.3 The curvature and type of expansion for universes containing both matter and a cosmological constant. The dashed line indicates $\kappa = 0$; models lying above this line have $\kappa = +1$, and those lying below have $\kappa = -1$. Also shown are the regions where the universe has a "Big Chill" expansion ($a \to \infty$ as $t \to \infty$), a "Big Crunch" recollapse ($a \to 0$ as $t \to t_{\text{crunch}}$), a loitering phase ($a \sim$ const for an extended period), or a "Big Bounce" ($a = a_{\min} > 0$ at $t = t_{\text{bounce}}$).

[5] A loitering universe is sometimes referred to as a *Lemaître universe*.

negatively curved, or flat. In the region labeled "Big Chill," the universe starts with $a = 0$ at $t = 0$, then expands outward forever, with $a \to \infty$ as $t \to \infty$. Like Big Crunch universes, Big Chill universes can have any sign for their curvature. In the region labeled "Big Bounce," the universe starts in a contracting state, reaches a minimum scale factor $a = a_{\text{min}} > 0$ at some time t_{bounce}, then expands outward forever, with $a \to \infty$ as $t \to \infty$. Universes that fall just below the dividing line between Big Bounce universes and Big Chill universes are loitering universes. The closer such a universe lies to the Big Bounce–Big Chill dividing line in Figure 6.3, the longer its loitering stage lasts.

To illustrate the different types of expansion and contraction possible, Figure 6.4 shows $a(t)$ for a set of four model universes. Each of these universes has the same current density parameter for matter: $\Omega_{m,0} = 0.3$, measured at $t = t_0$ and $a = 1$. These universes cannot be distinguished from each other by measuring their current matter density and Hubble constant. Nevertheless, thanks to their different values for the cosmological constant, they have very different pasts and very different futures. The dashed line in Figure 6.4 shows the scale factor for a universe with $\Omega_{\Lambda,0} = -0.3$; this universe has negative curvature, and is destined to end in a Big Crunch. The dotted line shows $a(t)$ for a universe with $\Omega_{\Lambda,0} = 0.7$; this universe is spatially flat, and is destined to end in an exponentially expanding Big Chill. The dot-dash line shows the scale factor for a universe with $\Omega_{\Lambda,0} = 1.7134$; this is a positively curved loitering universe, which spends

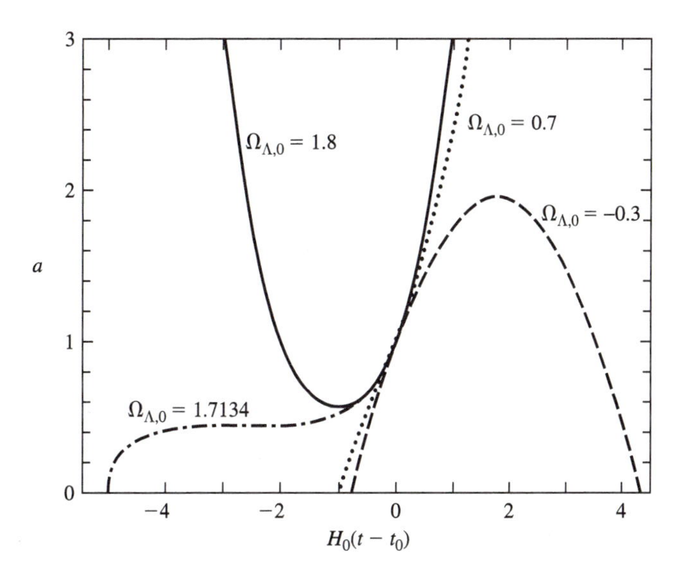

FIGURE 6.4 The scale factor a as a function of t in four different universes, each with $\Omega_{m,0} = 0.3$. The dashed line shows a "Big Crunch" universe ($\Omega_{\Lambda,0} = -0.3$, $\kappa = -1$). The dotted line shows a "Big Chill" universe ($\Omega_{\Lambda,0} = 0.7$, $\kappa = 0$). The dot-dash line shows a loitering universe ($\Omega_{\Lambda,0} = 1.7134$, $\kappa = +1$). The solid line shows a "Big Bounce" universe ($\Omega_{\Lambda,0} = 1.8$, $\kappa = +1$).

a long time with a scale factor $a \approx a_{\rm loiter} \approx 0.44$. Finally, the solid line shows a universe with $\Omega_{\Lambda,0} = 1.8$. This universe lies above the Big Chill–Big Bounce dividing line in Figure 6.3; it is a positively curved universe that "bounced" at a scale factor $a = a_{\rm bounce} \approx 0.56$.

There is strong observational evidence that we do not live in a loitering or Big Bounce universe. If we lived in a loitering universe, then as we looked out into space, we would see nearly the same redshift $z_{\rm loiter} = 1/a_{\rm loiter} - 1$ for galaxies with a very large range of distances. For instance, with $a_{\rm loiter} \approx 0.44$ (the appropriate loitering scale factor for a universe with $\Omega_{m,0} = 0.3$), this would lead to a large excess of galaxies with $z_{\rm loiter} \approx 1.3$. No such excess of galaxies is seen at any redshift in our universe. If we lived in a Big Bounce universe, then the largest redshift we would see for any galaxy would be $z_{\rm max} = 1/a_{\rm bounce} - 1$. As we looked further into space, we would see redshifts increase to $z_{\rm max}$, then see the redshifts decrease until they actually became blueshifts. In our universe, we do not see such distant blueshifted galaxies. Our own universe seems to be a Big Chill universe, fated to eternal expansion.

6.4 ■ RADIATION + MATTER

In our universe, radiation-matter equality took place at a scale factor $a_{rm} \equiv \Omega_{r,0}/\Omega_{m,0} \approx 2.8 \times 10^{-4}$. At scale factors $a \ll a_{rm}$, the universe is well described by a flat, radiation-only model, as described in section 5.5. At scale factors $a \sim a_{rm}$, the universe is better described by a flat model containing both radiation and matter. The Friedmann equation around the time of radiation-matter equality can be written in the approximate form

$$\frac{H^2}{H_0^2} = \frac{\Omega_{r,0}}{a^4} + \frac{\Omega_{m,0}}{a^3}. \tag{6.35}$$

This can be rearranged in the form

$$H_0 dt = \frac{a da}{\Omega_{r,0}^{1/2}} \left[1 + \frac{a}{a_{rm}} \right]^{-1/2}. \tag{6.36}$$

Integration yields a fairly simple relation for t as a function of a during the epoch when only radiation and matter are significant:

$$H_0 t = \frac{4a_{rm}^2}{3\sqrt{\Omega_{r,0}}} \left[1 - \left(1 - \frac{a}{2a_{rm}}\right)\left(1 + \frac{a}{a_{rm}}\right)^{1/2} \right]. \tag{6.37}$$

In the limit $a \ll a_{rm}$, this gives the appropriate result for the radiation-dominated phase of evolution (compare equation (6.10)),

$$a \approx \left(2\sqrt{\Omega_{r,0}}\, H_0 t\right)^{1/2} \qquad [a \ll a_{rm}]. \tag{6.38}$$

In the limit $a \gg a_{rm}$ (but before curvature or Λ contributes significantly to the Friedmann equation), the approximate result for $a(t)$ becomes

$$a \approx \left(\frac{3}{2}\sqrt{\Omega_{m,0}}\, H_0 t\right)^{2/3} \qquad [a \gg a_{rm}]. \tag{6.39}$$

The time of radiation-matter equality, t_{rm}, can be found by setting $a = a_{rm}$ in equation (6.37):

$$t_{rm} = \frac{4}{3}\left(1 - \frac{1}{\sqrt{2}}\right)\frac{a_{rm}^2}{\sqrt{\Omega_{r,0}}} H_0^{-1} \approx 0.391 \frac{\Omega_{r,0}^{3/2}}{\Omega_{m,0}^2} H_0^{-1}. \tag{6.40}$$

For the Benchmark Model, with $\Omega_{r,0} = 8.4 \times 10^{-5}$, $\Omega_{m,0} = 0.3$, and $H_0^{-1} =$ 14 Gyr, the time of radiation-matter equality was

$$t_{rm} = 3.34 \times 10^{-6} H_0^{-1} = 47{,}000\,\text{yr}. \tag{6.41}$$

The epoch when the universe was radiation-dominated was only about 47 millennia long. This is sufficiently brief that it justifies our ignoring the effects of radiation when computing the age of the universe. The age $t_0 = 0.964 H_0^{-1} = 13.5\,\text{Gyr}$ that we computed in section 6.2 (ignoring radiation) would only be altered by a few parts per million if we included the effects of radiation. This minor correction is dwarfed by the 10% uncertainty in the value of H_0^{-1}.

6.5 ■ BENCHMARK MODEL

The Benchmark Model, which we have adopted as the best fit to the currently available observational data, is spatially flat, and contains radiation, matter, and a cosmological constant. Some of its properties are listed, for ready reference, in Table 6.2. The Hubble constant of the Benchmark Model is assumed to be $H_0 = 70\,\text{km s}^{-1}\,\text{Mpc}^{-1}$. The radiation in the Benchmark Model consists of photons and neutrinos. The photons are assumed to be provided solely by a Cosmic Microwave Background with current temperature $T_0 = 2.725\,\text{K}$ and density parameter $\Omega_{\gamma,0} = 5.0 \times 10^{-5}$. The energy density of the cosmic neutrino background is theoretically calculated to be 68% of that of the Cosmic Microwave Background, as long as neutrinos are relativistic. If a neutrino has a nonzero mass m_ν, equation (5.18) tells us that it defects from the "radiation" column to the "matter" column when the scale factor is $a \sim 5 \times 10^{-4}\,\text{eV}/(m_\nu c^2)$. The matter content of the Benchmark Model consists partly of baryonic matter (that is, matter composed of protons and neutrons, with associated electrons), and partly of nonbaryonic dark matter. As we'll see in future chapters, the evidence indicates that most of the matter in the universe is nonbaryonic dark matter. The baryonic material that we are familiar with from our everyday existence has a density parameter of $\Omega_{\text{bary},0} \approx 0.04$ today. The density parameter of the nonbaryonic dark

TABLE 6.2 Properties of the Benchmark Model

	List of Ingredients	
photons:	$\Omega_{\gamma,0} = 5.0 \times 10^{-5}$	
neutrinos:	$\Omega_{\nu,0} = 3.4 \times 10^{-5}$	
total radiation:	$\Omega_{r,0} = 8.4 \times 10^{-5}$	
baryonic matter:	$\Omega_{\text{bary},0} = 0.04$	
nonbaryonic dark matter:	$\Omega_{\text{dm},0} = 0.26$	
total matter:	$\Omega_{m,0} = 0.30$	
cosmological constant:	$\Omega_{\Lambda,0} \approx 0.70$	
	Important Epochs	
radiation-matter equality:	$a_{rm} = 2.8 \times 10^{-4}$	$t_{rm} = 4.7 \times 10^4$ yr
matter-lambda equality:	$a_{m\Lambda} = 0.75$	$t_{m\Lambda} = 9.8$ Gyr
Now:	$a_0 = 1$	$t_0 = 13.5$ Gyr

matter is roughly six times greater: $\Omega_{\text{dm},0} \approx 0.26$. The bulk of the energy density in the Benchmark Model, however, is not provided by radiation or matter, but by a cosmological constant, with $\Omega_{\Lambda,0} = 1 - \Omega_{m,0} - \Omega_{r,0} \approx 0.70$.

The Benchmark Model was first radiation-dominated, then matter-dominated, and is now entering into its lambda-dominated phase. As we've seen, radiation gave way to matter at a scale factor $a_{rm} = \Omega_{r,0}/\Omega_{m,0} = 2.8 \times 10^{-4}$, correspond-

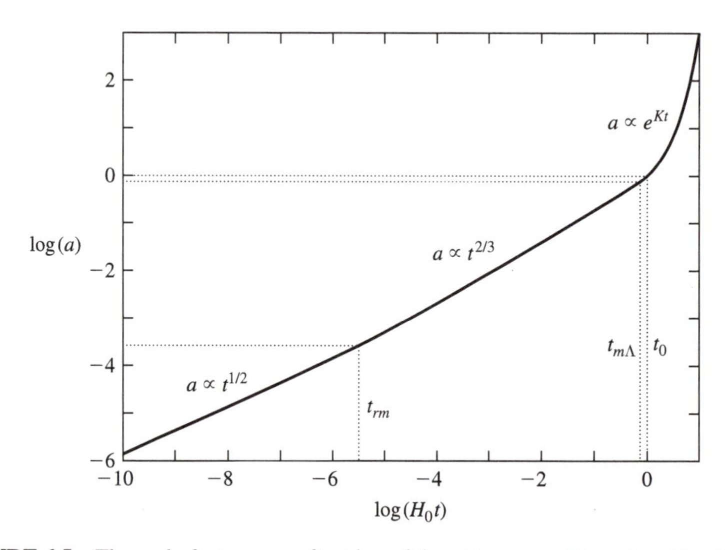

FIGURE 6.5 The scale factor a as a function of time t (measured in units of the Hubble time), computed for the Benchmark Model. The dotted lines indicate the time of radiation-matter equality, $a_{rm} = 2.8 \times 10^{-4}$, the time of matter-lambda equality, $a_{m\Lambda} = 0.75$, and the present moment, $a_0 = 1$.

ing to a time $t_{rm} = 4.7 \times 10^4$ yr. Matter, in turn, gave way to the cosmological constant at $a_{m\Lambda} = (\Omega_{m,0}/\Omega_{\Lambda,0})^{1/3} = 0.75$, corresponding to $t_{m\Lambda} = 9.8\,\text{Gyr}$. The current age of the universe, in the Benchmark Model, is $t_0 = 13.5\,\text{Gyr}$.

With $\Omega_{r,0}$, $\Omega_{m,0}$, and $\Omega_{\Lambda,0}$ known, the scale factor $a(t)$ can be computed numerically using the Friedmann equation, in the form of equation (6.6). Figure 6.5 shows the scale factor, thus computed, for the Benchmark Model. Note that the transition from the $a \propto t^{1/2}$ radiation-dominated phase to the $a \propto t^{2/3}$ matter-dominated phase is not an abrupt one; neither is the later transition from the matter-dominated phase to the exponentially growing lambda-dominated phase. One curious feature of the Benchmark Model illustrated vividly in Figure 6.5 is that we are living very close to the time of matter-lambda equality.

Once $a(t)$ is known, other properties of the Benchmark Model can be computed readily. For instance, the upper panel of Figure 6.6 shows the current proper

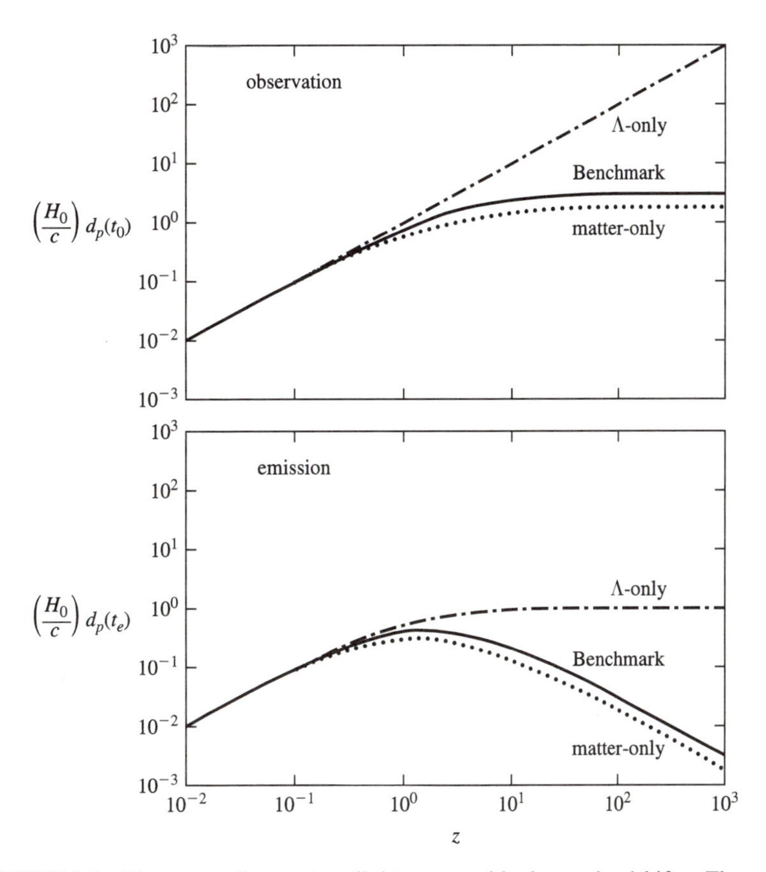

FIGURE 6.6 The proper distance to a light source with observed redshift z. The upper panel shows the distance at the time of observation; the lower panel shows the distance at the time of emission. The bold solid line indicates the Benchmark Model, the dot-dash line a flat, lambda-only universe, and the dotted line a flat, matter-only universe.

distance to a galaxy with redshift z. The heavy solid line is the result for the Benchmark Model; for purposes of comparison, the result for a flat lambda-only universe is shown as a dot-dash line and the result for a flat matter-only universe is shown as the dotted line. In the limit $z \to \infty$, the proper distance $d_p(t_0)$ approaches a limiting value $d_p \to 3.24c/H_0$, in the case of the Benchmark Model. Thus, the Benchmark Model has a finite horizon distance,

$$d_{\text{hor}}(t_0) = 3.24c/H_0 = 3.12ct_0 = 14{,}000\,\text{Mpc}. \tag{6.42}$$

If the Benchmark Model is a good description of our own universe, then we can't see objects more than 14 gigaparsecs away because light from them has not yet had time to reach us. The lower panel of Figure 6.6 shows $d_p(t_e)$, the distance to a galaxy with observed redshift z at the time the observed photons were emitted. For the Benchmark Model, $d_p(t_e)$ has a maximum for galaxies with redshift $z = 1.6$, where $d_p(t_e) = 0.41c/H_0$.

When astronomers observe a distant galaxy, they ask the related, but not identical, questions, "How far away is that galaxy?" and "How long has the light from that galaxy been traveling?" In the Benchmark Model, or any other model, we can answer the question "How far away is that galaxy?" by computing the proper distance $d_p(t_0)$. We can answer the question "How long has the light from that galaxy been traveling?" by computing the *lookback time*. If light emitted at time t_e is observed at time t_0, the lookback time is simply $t_0 - t_e$. In the limits of very small redshifts, $t_0 - t_e \approx z/H_0$. However, as shown in Figure 6.7, at larger redshifts the relation between lookback time and redshift becomes nonlin-

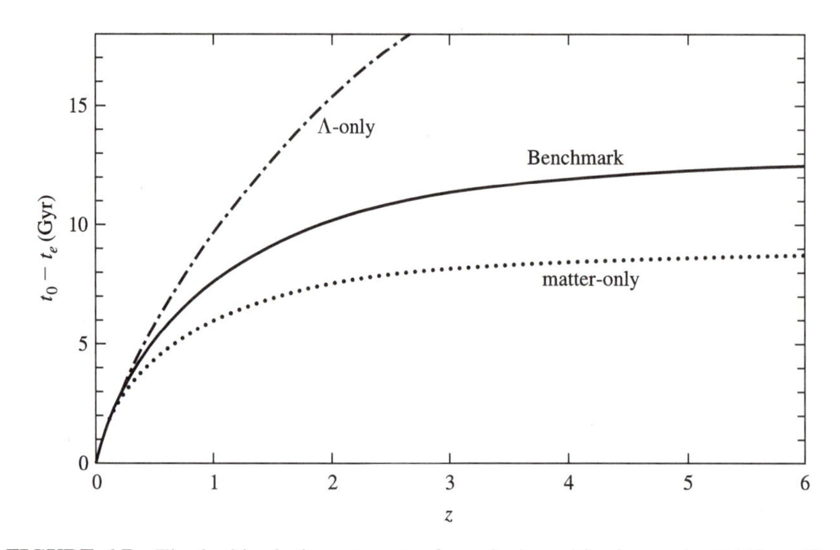

FIGURE 6.7 The lookback time, $t_0 - t_e$, for galaxies with observed redshift z. The Hubble time is assumed to be $H_0^{-1} = 14\,\text{Gyr}$. The heavy solid line shows the result for the Benchmark Model, the dot-dash line for a flat, lambda-only universe, and the dotted line for a flat, matter-only universe.

ear. The exact dependence of lookback time on redshift depends strongly on the cosmological model used. For example, consider a galaxy with redshift $z = 2$. In the Benchmark Model, the lookback time to that galaxy is 10.5 Gyr; we are seeing a redshifted image of that galaxy as it was 10.5 billion years ago. In a flat, lambda-only universe, however, the lookback time to a $z = 2$ galaxy is 15.4 Gyr, assuming $H_0^{-1} = 14\,\text{Gyr}$. In a flat, matter-dominated universe, the lookback time to a $z = 2$ galaxy is a mere 7.5 Gyr, with the same assumed Hubble constant. Knowing $\Omega_{m,0}$ and $\Omega_{\Lambda,0}$ thus becomes important to studies of galaxy evolution. How long does it take galaxies at $z \approx 2$ to evolve into galaxies similar to those at $z \approx 0$? Is it 15 billion years, or only half that time, or something in between? In future years, as the Benchmark Model becomes better constrained, our ability to translate observed redshifts into deduced times will become more accurate. The most distant galaxies that have been observed (at the beginning of the 21st century) are at a redshift $z \approx 6$. Consider such a high-redshift galaxy. Using the Benchmark Model, we find that the current proper distance to a galaxy with $z = 6$ is $d_p(t_0) = 1.92c/H_0 = 8300\,\text{Mpc}$, about 60% of the current horizon distance. The proper distance at the time the light was emitted was $d_p(t_e) = d_p(t_0)/(1 + z) = 0.27c/H_0 = 1200\,\text{Mpc}$. The light we observe now was emitted when the age of the universe was $t_e = 0.066 H_0^{-1} = 0.9\,\text{Gyr}$, or less than 7% of the universe's current age, $t_0 = 13.5\,\text{Gyr}$. The lookback time to a $z = 6$ galaxy in the Benchmark Model is thus $t_0 - t_e = 12.6\,\text{Gyr}$. Astronomers are fond of saying, "A telescope is a time machine."[6] As you look further and further out into the universe, to objects with larger and larger values of $d_p(t_0)$, you are looking back to objects with smaller and smaller values of t_e. When you observe a galaxy with a redshift $z = 6$, according to the Benchmark Model, you are glimpsing the universe as it was as a youngster, less than a billion years old.

SUGGESTED READING

Full references are given in the Annotated Bibliography on page 235.

Harrison (2000), ch. 18: A classification of possible universes, by kinematic and dynamic criteria

Kolb & Turner (1990), ch. 3.2: The scale factor $a(t)$ and current age t_0 for various two-component universes; contains useful formulae and graphs

PROBLEMS

6.10. In a positively curved universe containing only matter ($\Omega_0 > 1$, $\kappa = +1$), show that the present age of the universe is given by the formula

[6] Or, as William Herschel phrased it over two centuries ago, "A telescope with a power of penetrating into space...has also, as it may be called, a power of penetrating into time past."

$$H_0 t_0 = \frac{\Omega_0}{2(\Omega_0 - 1)^{3/2}} \cos^{-1}\left(\frac{2 - \Omega_0}{\Omega_0}\right) - \frac{1}{\Omega_0 - 1}. \quad (6.43)$$

Assuming $H_0 = 70\,\text{km s}^{-1}\,\text{Mpc}^{-1}$, plot t_0 as a function of Ω_0 in the range $1 \leq \Omega_0 \leq 3$.

6.11. In a negatively curved universe containing only matter ($\Omega_0 < 1$, $\kappa = -1$), show that the present age of the universe is given by the formula

$$H_0 t_0 = \frac{1}{1 - \Omega_0} - \frac{\Omega_0}{2(1 - \Omega_0)^{3/2}} \cosh^{-1}\left(\frac{2 - \Omega_0}{\Omega_0}\right). \quad (6.44)$$

Assuming $H_0 = 70\,\text{km s}^{-1}\,\text{Mpc}^{-1}$, plot t_0 as a function of Ω_0 in the range $0 \leq \Omega_0 \leq 1$. The current best estimate for the ages of stars in globular clusters yields an age of $t = 13\,\text{Gyr}$ for the oldest globular clusters. In a matter-only universe, what is the maximum permissible value of Ω_0, given the constraints $H_0 = 70\,\text{km s}^{-1}\,\text{Mpc}^{-1}$ and $t_0 > 13\,\text{Gyr}$?

6.12. One of the more recent speculations in cosmology is that the universe may contain a quantum field, called "quintessence," which has a positive energy density and a negative value of the equation-of-state parameter w. Assume, for the purposes of this problem, that the universe is spatially flat, and contains nothing but matter ($w = 0$), and quintessence with $w = -\frac{1}{2}$. The current density parameter of matter is $\Omega_{m,0} \leq 1$, and the current density parameter of quintessence is $\Omega_{Q,0} = 1 - \Omega_{m,0}$. At what scale factor a_{mQ} will the energy density of quintessence and matter be equal? Solve the Friedmann equation to find $a(t)$ for this universe. What is $a(t)$ in the limit $a \ll a_{mQ}$? What is $a(t)$ in the limit $a \gg a_{mQ}$? What is the current age of this universe, expressed in terms of H_0 and $\Omega_{m,0}$?

6.13. Suppose you wanted to "pull an Einstein," and create a static universe ($\dot{a} = 0$, $\ddot{a} = 0$) in which the gravitational attraction of matter is exactly balanced by the gravitational repulsion of quintessence with equation-of-state parameter w_Q. Within what range must w_Q fall for the effects of quintessence to be repulsive? For repulsive quintessence with energy density ε_Q and equation-of-state parameter w_Q, what is the necessary matter density (ε_m) to produce a static universe? Will the curvature of this static universe be negative or positive? What will be its radius of curvature, expressed in terms of ε_Q and w_Q?

6.14. Consider a positively curved universe containing only matter (the "Big Crunch" model discussed in section 6.1). At some time $t_0 > t_{\text{Crunch}}/2$, during the contraction phase of this universe, an astronomer named Elbbuh Niwde discovers that nearby galaxies have blueshifts ($-1 \leq z < 0$) proportional to their distance. He then measures H_0 and Ω_0, finding $H_0 < 0$ and $\Omega_0 > 1$. Given H_0 and Ω_0, how long a time will elapse between Dr. Niwde's observations at $t = t_0$ and the final Big Crunch at $t = t_{\text{crunch}}$? What is the minimum blueshift that Dr. Niwde is able to observe? What is the lookback time to an object with this blueshift?

6.15. Consider an expanding, positively curved universe containing only a cosmological constant ($\Omega_0 = \Omega_{\Lambda,0} > 1$). Show that such a universe underwent a "Big Bounce" at a scale factor

$$a_{\text{bounce}} = \left(\frac{\Omega_0 - 1}{\Omega_0} \right)^{1/2}, \tag{6.45}$$

and that the scale factor as a function of time is

$$a(t) = a_{\text{bounce}} \cosh[\sqrt{\Omega_0} H_0 (t - t_{\text{bounce}})], \tag{6.46}$$

where t_{bounce} is the time at which the Big Bounce occurred. What is the time $t_0 - t_{\text{bounce}}$ that has elapsed since the Big Bounce, expressed as a function of H_0 and Ω_0?

6.16. A universe is spatially flat, and contains both matter and a cosmological constant. For what value of $\Omega_{m,0}$ is t_0 exactly equal to H_0^{-1}?

6.17. In the Benchmark Model, what is the total mass of all the matter within our horizon? What is the total energy of all the photons within our horizon? How many baryons are within the horizon?

CHAPTER 7

Measuring Cosmological Parameters

Cosmologists would like to know the scale factor $a(t)$ for the universe. For a model universe whose contents are known with precision, the scale factor can be computed from the Friedmann equation. Finding $a(t)$ for the real universe, however, is much more difficult. The scale factor is not directly observable; it can only be deduced indirectly from the imperfect and incomplete observations that we make of the universe around us.

In the previous three chapters, I've pointed out that if we knew the energy density ε for each component of the universe, we could use the Friedmann equation to find the scale factor $a(t)$. The argument works in the other direction, as well; if we could determine $a(t)$ from observations, we could use that knowledge to find ε for each component. Let's see, then, what constraints we can put on the scale factor by making observations of distant astronomical objects.

7.1 ■ "A SEARCH FOR TWO NUMBERS"

Since determining the exact functional form of $a(t)$ is difficult, it is useful, instead, to do a Taylor series expansion for $a(t)$ around the present moment. The complete Taylor series is

$$a(t) = a(t_0) + \left.\frac{da}{dt}\right|_{t=t_0} (t - t_0) + \frac{1}{2} \left.\frac{d^2a}{dt^2}\right|_{t=t_0} (t - t_0)^2 + \cdots \tag{7.1}$$

To exactly reproduce an arbitrary function $a(t)$ for all values of t, an infinite number of terms is required in the expansion. However, the usefulness of a Taylor series expansion resides in the fact that if a doesn't fluctuate wildly with t, using only the first few terms of the expansion gives a good approximation in the immediate vicinity of t_0. The scale factor $a(t)$ is a good candidate for a Taylor expansion. The different model universes examined in the previous two chapters all had smoothly varying scale factors, and there's no evidence that the real universe has a wildly oscillating scale factor.

Keeping the first three terms of the Taylor expansion, the scale factor in the recent past and the near future can be approximated as

$$a(t) \approx a(t_0) + \left.\frac{da}{dt}\right|_{t=t_0} (t - t_0) + \frac{1}{2} \left.\frac{d^2 a}{dt^2}\right|_{t=t_0} (t - t_0)^2. \quad (7.2)$$

Dividing by the current scale factor, $a(t_0)$,

$$\frac{a(t)}{a(t_0)} \approx 1 + \left.\frac{\dot{a}}{a}\right|_{t=t_0} (t - t_0) + \frac{1}{2} \left.\frac{\ddot{a}}{a}\right|_{t=t_0} (t - t_0)^2. \quad (7.3)$$

Using the normalization $a(t_0) = 1$, this expansion for the scale factor is customarily written in the form

$$a(t) \approx 1 + H_0(t - t_0) - \frac{1}{2} q_0 H_0^2 (t - t_0)^2. \quad (7.4)$$

In equation (7.4), the parameter H_0 is our old acquaintance the Hubble constant,

$$H_0 \equiv \left.\frac{\dot{a}}{a}\right|_{t=t_0}, \quad (7.5)$$

and the parameter q_0 is a dimensionless number called the *deceleration parameter*, defined as

$$q_0 \equiv -\left(\frac{\ddot{a}a}{\dot{a}^2}\right)_{t=t_0} = -\left(\frac{\ddot{a}}{aH^2}\right)_{t=t_0}. \quad (7.6)$$

Note the choice of sign in defining q_0. A positive value of q_0 corresponds to $\ddot{a} < 0$, meaning that the universe's expansion is decelerating (that is, the relative velocity of any two points is decreasing). A negative value of q_0 corresponds to $\ddot{a} > 0$, meaning that the relative velocity of any two points is increasing with time. The choice of sign for q_0, and the fact that it's named the *deceleration* parameter, is because it was first defined during the mid-1950's, when the limited information available favored a matter-dominated universe with $\ddot{a} < 0$. If the universe contains a sufficiently large cosmological constant, however, the deceleration parameter q_0 can have either sign.

The Taylor expansion of equation (7.4) is physics-free. It is simply a mathematical description of how the universe expands at times $t \sim t_0$, and says nothing at all about what forces act to accelerate the expansion (to take a Newtonian viewpoint of the physics involved). The parameters H_0 and q_0 are thus purely descriptive of the kinematics, and are free of the theoretical "baggage" underlying the Friedmann equation and the acceleration equation.[1] In a famous 1970 review

[1]Remember, the Friedmann equation assumes that the expansion of the universe is controlled by gravity, and that gravity is accurately described by Einstein's theory of general relativity; although these are reasonable assumptions, they are not 100% iron-clad.

article, the observational cosmologist Allan Sandage described all of cosmology as "A Search for Two Numbers." Those two numbers were H_0 and q_0. Although the scope of cosmology has widened considerably since Sandage wrote his article, cosmologists are still assiduously searching for H_0 and q_0.

Although H_0 and q_0 are themselves free of the theoretical assumptions underlying the Friedmann and acceleration equations, we can use the acceleration equation to predict what q_0 will be in a given model universe. If our model universe contains several components, each with a different value of the equation-of-state parameter w, the acceleration equation can be written

$$\frac{\ddot{a}}{a} = -\frac{4\pi G}{3c^2} \sum_w \varepsilon_w (1 + 3w). \tag{7.7}$$

Divide each side of the acceleration equation by the square of the Hubble parameter $H(t)$ and change sign:

$$-\frac{\ddot{a}}{aH^2} = \frac{1}{2} \left[\frac{8\pi G}{3c^2 H^2} \right] \sum_w \varepsilon_w (1 + 3w). \tag{7.8}$$

However, the quantity in square brackets in equation (7.8) is just the inverse of the critical energy density ε_c. Thus, we can rewrite the acceleration equation in the form

$$-\frac{\ddot{a}}{aH^2} = \frac{1}{2} \sum_w \Omega_w (1 + 3w). \tag{7.9}$$

Evaluating equation (7.9) at the present moment, $t = t_0$, tells us the relation between the deceleration parameter q_0 and the density parameters of the different components of the universe:

$$q_0 = \frac{1}{2} \sum_w \Omega_{w,0} (1 + 3w). \tag{7.10}$$

For a universe containing radiation, matter, and a cosmological constant,

$$q_0 = \Omega_{r,0} + \frac{1}{2}\Omega_{m,0} - \Omega_{\Lambda,0}. \tag{7.11}$$

Such a universe will currently be accelerating outward ($q_0 < 0$) if $\Omega_{\Lambda,0} > \Omega_{r,0} + \Omega_{m,0}/2$. The Benchmark Model, for instance, has $q_0 \approx -0.55$.

In principle, determining H_0 should be easy. For small redshifts, the relation between a galaxy's distance d and its redshift z is linear (equation (2.5)):

$$cz = H_0 d. \tag{7.12}$$

Thus, if you measure the distance d and redshift z for a large sample of galaxies, and fit a straight line to a plot of cz versus d, the slope of the plot gives you the value of H_0.[2] Measuring the redshift of a galaxy is relatively simple; automated galaxy surveys can find hundreds of galaxy redshifts in a single night. The difficulty is in measuring the *distance* of a galaxy. Remember, Edwin Hubble was off by a factor of 7 when he estimated $H_0 \approx 500\,\text{km}\,\text{s}^{-1}\,\text{Mpc}^{-1}$ (see Figure 2.4). This is because he underestimated the distances to galaxies in his sample by a factor of 7.

The distance to a galaxy is not only difficult to measure, but also, in an expanding universe, somewhat difficult to define. In section 3.3, the proper distance $d_p(t)$ between two points was defined as the length of the spatial geodesic between the points when the scale factor is fixed at the value $a(t)$. The proper distance is perhaps the most straightforward definition of the spatial distance between two points in an expanding universe. Moreover, there is a helpful relation between scale factor and proper distance. If we observe, at time t_0, light that was emitted by a distant galaxy at time t_e, the current proper distance to that galaxy is (equation (5.35)):

$$d_p(t_0) = c \int_{t_e}^{t_0} \frac{dt}{a(t)}. \tag{7.13}$$

For the model universes examined in Chapters 5 and 6, we knew the exact functional form of $a(t)$, and hence could exactly compute $d_p(t_0)$ for a galaxy of any redshift. If we have only partial knowledge of the scale factor, in the form of the Taylor expansion of equation (7.4), we may use the expansion

$$\frac{1}{a(t)} \approx 1 - H_0(t - t_0) + \left(\frac{1 + q_0}{2}\right) H_0^2 (t - t_0)^2 \tag{7.14}$$

in equation (7.13). Including the two lowest-order terms in the lookback time, $t_0 - t_e$, we find that the proper distance to the galaxy is

$$d_p(t_0) \approx c(t_0 - t_e) + \frac{cH_0}{2}(t_0 - t_e)^2. \tag{7.15}$$

The first term in the above equation, $c(t_0 - t_e)$, is what the proper distance would be in a static universe—the lookback time times the speed of light. The second term is a correction due to the expansion of the universe during the time the light was traveling.

Equation (7.15) would be extremely useful if the photons from distant galaxies carried a stamp telling us the lookback time, $t_0 - t_e$. They don't; instead, they carry a stamp telling us the scale factor $a(t_e)$ at the time the light was emitted. The observed redshift z of a galaxy, remember, is

$$z = \frac{1}{a(t_e)} - 1. \tag{7.16}$$

[2]The peculiar velocities of galaxies cause a significant amount of scatter in the plot, but by using a large number of galaxies, you can beat down the statistical errors. If you use galaxies at $d < 100\,\text{Mpc}$, you must also make allowances for the local inhomogeneity and anisotropy.

Using equation (7.14), we may write an approximate relation between redshift and lookback time:

$$z \approx H_0(t_0 - t_e) + \left(\frac{1+q_0}{2}\right) H_0^2 (t_0 - t_e)^2. \tag{7.17}$$

Inverting equation (7.17) to give the lookback time as a function of redshift, we find

$$t_0 - t_e \approx H_0^{-1}\left[z - \left(\frac{1+q_0}{2}\right) z^2\right]. \tag{7.18}$$

Substituting equation (7.18) into equation (7.15) gives us an approximate relation for the current proper distance to a galaxy with redshift z:

$$d_p(t_0) \approx \frac{c}{H_0}\left[z - \left(\frac{1+q_0}{2}\right) z^2\right] + \frac{cH_0}{2}\frac{z^2}{H_0^2} = \frac{c}{H_0} z \left[1 - \frac{1+q_0}{2} z\right]. \tag{7.19}$$

The linear Hubble relation $d_p \propto z$ thus holds true only in the limit $z \ll 2/(1+q_0)$. If $q_0 > -1$, then the proper distance to a galaxy of moderate redshift ($z \sim 0.1$, say) is less than would be predicted from the linear Hubble relation.

7.2 ■ LUMINOSITY DISTANCE

Unfortunately, the current proper distance to a galaxy, $d_p(t_0)$, is not a measurable property. If you tried to measure the distance to a galaxy with a tape measure, for instance, the distance would be continuously increasing as you extended the tape. To measure the proper distance at time t_0, you would need a tape measure that could be extended with infinite speed; alternatively, you would need to stop the expansion of the universe at its current scale factor while you measured the distance at your leisure. Neither of these alternatives is physically possible.

Since cosmology is ultimately based on observations, if we want to find the distance to a galaxy, we need some way of computing a distance from that galaxy's observed properties. In devising ways of computing the distance to galaxies, astronomers have found it useful to adopt and adapt the techniques used to measure shorter distances. Let's examine, then, the techniques used to measure relatively short distances. Within the solar system, astronomers measure the distance to the Moon and planets by reflecting radar signals from them. If δt is the time taken for a photon to complete the round-trip, then the distance to the reflecting body is $d = (c\,\delta t)/2$.[3] The accuracy with which distances have been determined with this technique is impressive; the length of the astronomical unit, for instance, is now known to be 1 AU = 149,597,870.61 km. The radar technique is useful only within the solar system. Beyond $\sim$ 10 AU, the reflected radio waves are too faint to detect.

[3]Since the relative speeds of objects within the solar system are much smaller than c, the corrections due to relative motion during the time δt are minuscule.

A favorite method for determining distances to other stars within our galaxy is the method of trigonometric parallax. When a star is observed from two points separated by a distance b, the star's apparent position will shift by an angle θ. If the baseline of observation is perpendicular to the line of sight to the star, the *parallax distance* will be

$$d_\pi = 1\,\text{pc}\left(\frac{b}{1\,\text{AU}}\right)\left(\frac{\theta}{1\,\text{arcsec}}\right)^{-1}. \tag{7.20}$$

Measuring the distances to stars using the Earth's orbit ($b = 2\,\text{AU}$) as a baseline is a standard technique. Since the size of the Earth's orbit is known with great accuracy from radar measurements, the accuracy with which the parallax distance can be determined is limited by the accuracy with which θ can be measured. The Hipparcos satellite, launched by the European Space Agency in 1989, found the parallax distance for $\sim 10^5$ stars, with an accuracy of ~ 1 milliarcsecond. However, to measure θ for a galaxy $\sim 100\,\text{Mpc}$ away, an accuracy of < 10 nanoarcseconds would be required, using the Earth's orbit as a baseline. The trigonometric parallaxes of galaxies at cosmological distances are too small to be measured with current technology.

Let's focus on the properties that we *can* measure for objects at cosmological distances. We can measure the flux of light, f, from the object, in units of watts per square meter. The complete flux, integrated over all wavelengths of light, is called the *bolometric* flux. (A bolometer is an extremely sensitive thermometer capable of detecting electromagnetic radiation over a wide range of wavelengths; it was invented in 1881 by the astronomer Samuel Langley who used it to measure solar radiation.[4]) More frequently, given the difficulties of measuring the true bolometric flux, the flux over a limited range of wavelengths is measured. If the light from the object has emission or absorption lines, we can measure the redshift, z. If the object is an extended source rather than a point of light, we can measure its angular diameter, $\delta\theta$.

One way of using measured properties to assign a distance is the *standard candle* method. A standard candle is an object whose luminosity L is known. For instance, if some class of astronomical object had luminosities that were the same throughout all of space-time, they would act as excellent standard candles—if their unique luminosity L were known. If you know, by some means or other, the luminosity of an object, then you can use its measured flux f to define a function called the *luminosity distance*:

$$d_L \equiv \left(\frac{L}{4\pi f}\right)^{1/2}. \tag{7.21}$$

The function d_L is called a "distance" because its dimensionality is that of a distance, and because it is what the proper distance to the standard candle would

[4]As expressed more poetically in an anonymous limerick: "Oh, Langley devised the bolometer: / It's really a kind of thermometer / Which measures the heat / From a polar bear's feet / At a distance of half a kilometer."

be *if* the universe were static and Euclidean. In a static Euclidean universe, the propagation of light follows the inverse square law $f = L/[4\pi d^2]$.

Suppose, though, that you are in a universe described by a Robertson–Walker metric (equation (3.25)):

$$ds^2 = -c^2 dt^2 + a(t)^2 [dr^2 + S_\kappa(r)^2 d\Omega^2], \tag{7.22}$$

with

$$S_\kappa(r) = \begin{cases} R_0 \sin(r/R_0) & (\kappa = +1) \\ r & (\kappa = 0) \\ R_0 \sinh(r/R_0) & (\kappa = -1). \end{cases} \tag{7.23}$$

You are at the origin. At the present moment, $t = t_0$, you see light that was emitted by a standard candle at comoving coordinate location (r, θ, ϕ) at a time t_e (see Figure 7.1). The photons emitted at time t_e are, at the present moment, spread over a sphere of proper radius $d_p(t_0) = r$ and proper surface area $A_p(t_0)$. If space is flat ($\kappa = 0$), then the proper area of the sphere is given by the Euclidean relation $A_p(t_0) = 4\pi d_p(t_0)^2 = 4\pi r^2$. More generally, however,

$$A_p(t_0) = 4\pi S_\kappa(r)^2. \tag{7.24}$$

When space is positively curved, $A_p(t_0) < 4\pi r^2$, and the photons are spread over a *smaller* area than they would be in flat space. When space is negatively curved, $A_p(t_0) > 4\pi r^2$, and photons are spread over a *larger* area than they would be in flat space.

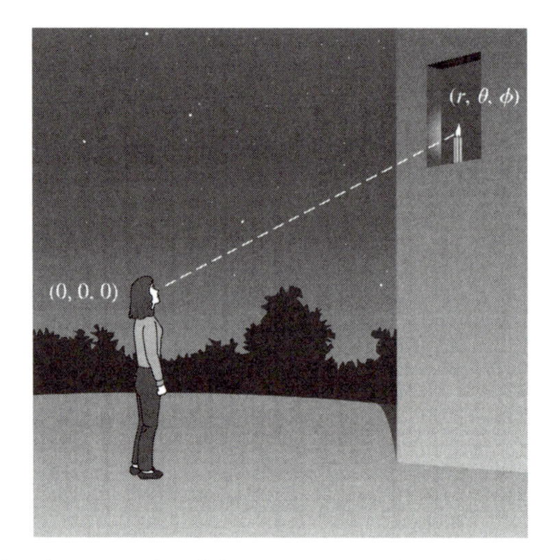

FIGURE 7.1 An observer at the origin observes a standard candle, of known luminosity L, at comoving coordinate location (r, θ, ϕ).

In addition to these geometric effects, which would apply even in a static universe, the expansion of the universe causes the observed flux of light from a standard candle of redshift z to be decreased by a factor of $(1+z)^{-2}$. First, the expansion of the universe causes the energy of each photon from the standard candle to decrease. If a photon starts with an energy $E_e = hc/\lambda_e$ when the scale factor is $a(t_e)$, by the time we observe it, when the scale factor is $a(t_0) = 1$, the wavelength will have grown to

$$\lambda_0 = \frac{1}{a(t_e)}\lambda_e = (1+z)\lambda_e, \tag{7.25}$$

and the energy will have fallen to

$$E_0 = \frac{E_e}{1+z}. \tag{7.26}$$

Second, thanks to the expansion of the universe, the time between photon detections will be greater. If two photons are emitted in the same direction separated by a time interval δt_e, the proper distance between them will initially be $c(\delta t_e)$; by the time we detect the photons at time t_0, the proper distance between them will be stretched to $c(\delta t_e)(1+z)$, and we will detect them separated by a time interval $\delta t_0 = \delta t_e(1+z)$.

The net result is that in an expanding, spatially curved universe, the relation between the observed flux f and the luminosity L of a distant light source is

$$f = \frac{L}{4\pi S_\kappa(r)^2(1+z)^2}, \tag{7.27}$$

and the luminosity distance is

$$d_L = S_\kappa(r)(1+z). \tag{7.28}$$

The available evidence indicates that our universe is nearly flat, with a radius of curvature R_0 larger than the current horizon distance $d_{\rm hor}(t_0)$. Objects with finite redshift are at proper distances smaller than the horizon distance, and hence smaller than the radius of curvature. Thus, it is safe to make the approximation $r \ll R_0$, implying $S_\kappa(r) \approx r$. With our assumption that space is very close to being flat, the relation between the luminosity distance and the current proper distance becomes very simple:

$$d_L = r(1+z) = d_p(t_0)(1+z) \qquad [\kappa = 0]. \tag{7.29}$$

Thus, even if space is perfectly flat, if you estimate the distance to a standard candle by using a naïve inverse square law, you will overestimate the actual proper distance by a factor $(1+z)$, where z is the standard candle's redshift.

Figure 7.2 shows the luminosity distance d_L as a function of redshift for the Benchmark Model, and for two other flat universes, one dominated by matter and one dominated by a cosmological constant. When $z \ll 1$, the current proper

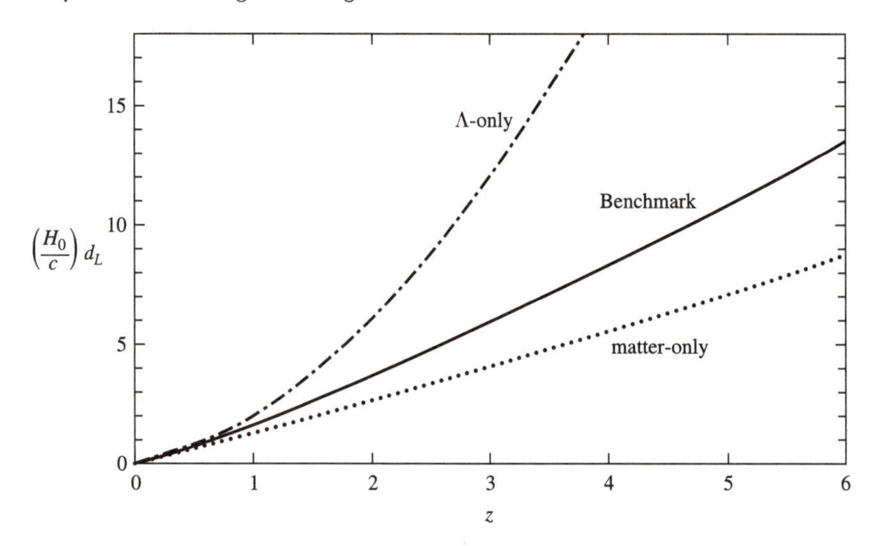

FIGURE 7.2 The luminosity distance of a standard candle with observed redshift z. The bold solid line gives the result for the Benchmark Model, the dot-dash line for a flat, lambda-only universe, and the dotted line for a flat, matter-only universe.

distance may be approximated as

$$d_P(t_0) \approx \frac{c}{H_0} z \left(1 - \frac{1+q_0}{2} z\right). \tag{7.30}$$

In a nearly flat universe, the luminosity distance may thus be approximated as

$$d_L \approx \frac{c}{H_0} z \left(1 - \frac{1+q_0}{2} z\right)(1+z) \approx \frac{c}{H_0} z \left(1 + \frac{1-q_0}{2} z\right). \tag{7.31}$$

Note that in the limit $z \to 0$,

$$d_p(t_0) \approx d_L \approx \frac{c}{H_0} z. \tag{7.32}$$

In a universe described by the Robertson–Walker metric, the luminosity distance is a good approximation to the current proper distance for objects with small redshifts.

7.3 ■ ANGULAR-DIAMETER DISTANCE

The luminosity distance d_L is not the only distance measure that can be computed using the observable properties of cosmological objects. Suppose that instead of a standard candle, you observed a *standard yardstick*. A standard yardstick is an object whose proper length ℓ is known. In most cases, it is convenient to choose as your yardstick an object that is tightly bound together, by gravity or duct tape

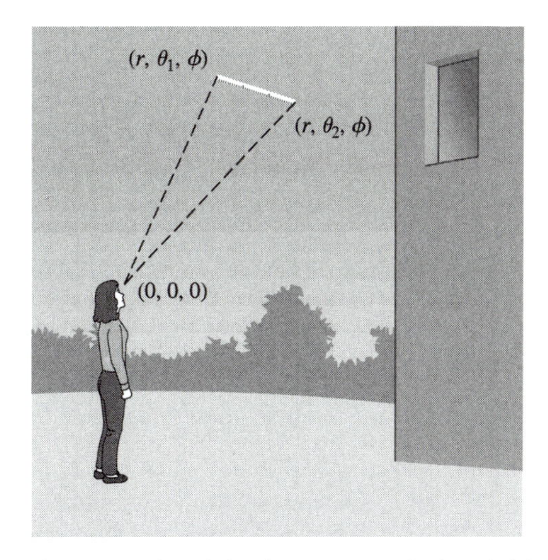

FIGURE 7.3 An observer at the origin observes a standard yardstick, of known proper length ℓ, at comoving coordinate distance r.

or some other influence, and hence is not expanding along with the universe as a whole.

Suppose a yardstick of constant proper length ℓ is aligned perpendicular to your line of sight, as shown in Figure 7.3. Measure an angular distance $\delta\theta$ between the ends of the yardstick, and a redshift z for the light that the yardstick emits. If $\delta\theta \ll 1$, and if you know the length ℓ of the yardstick, you can compute a distance to the yardstick using the small-angle formula

$$d_A \equiv \frac{\ell}{\delta\theta}. \tag{7.33}$$

This function of ℓ and $\delta\theta$ is called the *angular-diameter distance*. The angular-diameter distance is equal to the proper distance to the yardstick if the universe is static and Euclidean.

In general, though, if the universe is expanding or curved, the angular-diameter distance will not be equal to the current proper distance. Suppose you are in a universe described by the Robertson–Walker metric given in equation (7.22). Choose your comoving coordinate system so that you are at the origin. The yardstick is at a comoving coordinate distance r. At a time t_e, the yardstick emitted the light that you observe at time t_0. The comoving coordinates of the two ends of the yardstick, at the time the light was emitted, were (r, θ_1, ϕ) and (r, θ_2, ϕ). As the light from the yardstick moves toward the origin, it travels along geodesics with $\theta =$ constant and $\phi =$ constant. Thus, the angular size you measure for the yardstick will be $\delta\theta = \theta_2 - \theta_1$. The distance ds between the two ends of the yardstick, measured at the time t_e when the light was emitted, can be found from the Robertson–Walker metric:

$$ds = a(t_e)S_\kappa(r)\delta\theta. \tag{7.34}$$

However, for a standard yardstick whose length ℓ is known, we can set $ds = \ell$, and thus find that

$$\ell = a(t_e)S_\kappa(r)\delta\theta = \frac{S_\kappa(r)\delta\theta}{1+z}. \tag{7.35}$$

Thus, the angular-diameter distance d_A to a standard yardstick is

$$d_A \equiv \frac{\ell}{\delta\theta} = \frac{S_\kappa(r)}{1+z}. \tag{7.36}$$

Comparison with equation (7.28) shows that the relation between the angular-diameter distance and the luminosity distance is

$$d_A = \frac{d_L}{(1+z)^2}. \tag{7.37}$$

Thus, if you observe an object that is both a standard candle and a standard yardstick, the angular-diameter distance that you compute for the object will be smaller than the luminosity distance. Moreover, if the universe is spatially flat,

$$d_A(1+z) = d_p(t_0) = \frac{d_L}{1+z} \qquad [\kappa = 0]. \tag{7.38}$$

In a flat universe, therefore, if you compute the angular-diameter distance d_A of a standard yardstick, it isn't equal to the current proper distance $d_p(t_0)$; rather, it is equal to the proper distance at the time the light from the object was emitted: $d_A = d_p(t_0)/(1+z) = d_p(t_e)$.

Figure 7.4 shows the angular-diameter distance d_A for the Benchmark Model, and for two other spatially flat universes, one dominated by matter and one dominated by a cosmological constant. [Since d_A is, for these flat universes, equal to $d_p(t_e)$, Figure 7.4 is simply a replotting of the lower panel in Figure 6.6.] When $z \ll 1$, the approximate value of d_A is given by the expansion

$$d_A \approx \frac{c}{H_0} z \left(1 - \frac{3+q_0}{2} z\right). \tag{7.39}$$

Thus, comparing equations (7.30), (7.31), and (7.39), we find that in the limit $z \to 0$, $d_A \approx d_L \approx d_p(t_0) \approx (c/H_0)z$. However, the state of affairs is very different in the limit $z \to \infty$. In models with a finite horizon size, $d_p(t_0) \to d_{\rm hor}(t_0)$ as $z \to \infty$. The luminosity distance to highly redshifted objects, in this case, diverges as $z \to \infty$, with

$$d_L(z \to \infty) \approx z d_{\rm hor}(t_0). \tag{7.40}$$

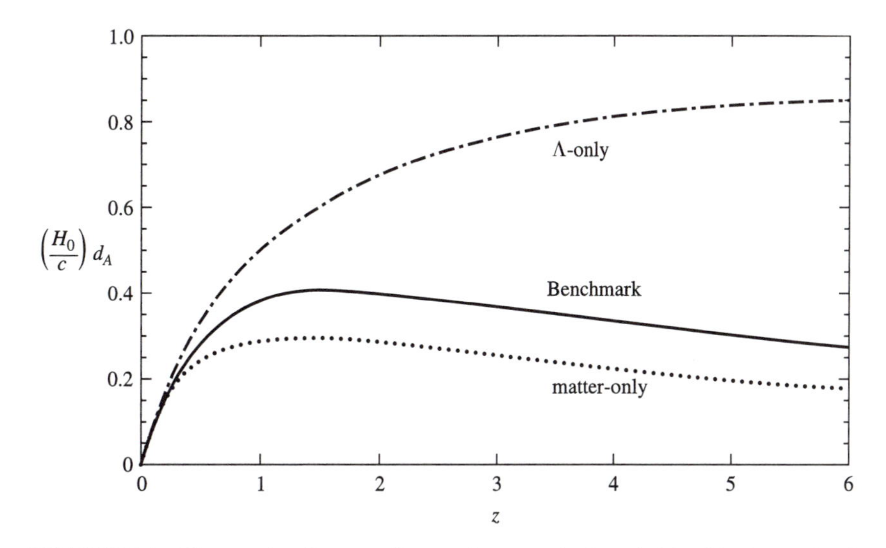

FIGURE 7.4 The angular-diameter distance for a standard yardstick with observed redshift z. The bold solid line gives the result for the Benchmark Model, the dot-dash line for a flat, lambda-only universe, and the dotted line for a flat, matter-only universe.

However, the angular-diameter distance to highly redshifted objects approaches zero as $z \to \infty$, with

$$d_A(z \to \infty) \approx \frac{d_{\rm hor}(t_0)}{z}. \tag{7.41}$$

In model universes other than the lambda-only model, the angular-diameter distance d_A has a maximum for standard yardsticks at some critical redshift z_c. (For the Benchmark Model, $z_c = 1.6$, where $d_A(\text{max}) = 0.41c/H_0 = 1800\,\text{Mpc}$.) This means that if the universe were full of glow-in-the-dark yardsticks, all of the same size ℓ, their angular size $\delta\theta$ would decrease with redshift out to $z = z_c$, but then would increase at larger redshifts. The sky would be full of big, faint, redshifted yardsticks.

In principle, standard yardsticks can be used to determine H_0. To begin with, identify a population of standard yardsticks (objects whose physical size ℓ is known). Then, measure the redshift z and angular size $\delta\theta$ of each standard yardstick. Compute the angular-diameter distance $d_A = \ell/\delta\theta$ for each standard yardstick. Plot cz versus d_A, and the slope of the relation, in the limit $z \to 0$, will give you H_0. In addition, if you have measured the angular size $\delta\theta$ for standard candles at $z \sim z_c$, the shape of the cz versus d_A plot can be used to determine further cosmological parameters. If you simply want a kinematic description, estimate q_0 by fitting equation (7.39) to the data. If you are confident that the universe is dominated by matter and a cosmological constant, you can see which values of $\Omega_{m,0}$ and $\Omega_{\Lambda,0}$ provide the best fit to the observed data.

In practice, the use of standard yardsticks to determine cosmological parameters has long been plagued with observational difficulties. For instance, a standard yardstick must have an angular size large enough to be resolved by your telescope. A yardstick of physical size ℓ will have its angular size $\delta\theta$ minimized when it is at the critical redshift z_c. For the Benchmark Model,

$$\delta\theta(\text{min}) = \frac{\ell}{d_A(\text{max})} = \frac{\ell}{1800\,\text{Mpc}} \approx 0.1\,\text{arcsec}\left(\frac{\ell}{1\,\text{kpc}}\right). \tag{7.42}$$

Both galaxies and clusters of galaxies are large enough to be useful standard candles. Unfortunately for cosmologists, galaxies and clusters of galaxies do not have sharply defined edges, so assigning a particular angular size $\delta\theta$, and a corresponding physical size ℓ, to these objects is a somewhat tricky task. Moreover, galaxies and clusters of galaxies are not isolated, rigid yardsticks of fixed length. Galaxies tend to become larger with time as they undergo mergers with their neighbors. Clusters, too, tend to become larger with time, as galaxies fall into them. (Eventually, our Local Group will fall into the Virgo cluster.) Correcting for these evolutionary trends is a difficult task.

Given the difficulties involved in using standard yardsticks to determine cosmological parameters, more attention has been focused, in recent years, on the use of standard candles. Let's first look, therefore, at how standard candles can be used to determine H_0, then focus on how they can be used to determine the acceleration of the universe.

7.4 ■ STANDARD CANDLES AND THE HUBBLE CONSTANT

Using standard candles to determine the Hubble constant has a long and honorable history; it's the method used by Hubble himself. The recipe for finding the Hubble constant is a simple one:

- Identify a population of standard candles with luminosity L.
- Measure the redshift z and flux f for each standard candle.
- Compute $d_L = (L/4\pi f)^{1/2}$ for each standard candle.
- Plot cz versus d_L.
- Measure the slope of the cz versus d_L relation when $z \ll 1$; this gives H_0.

As with the apocryphal recipe for rabbit stew that begins "First catch your rabbit," the hardest step is the first one. A good standard candle is hard to find. For cosmological purposes, a standard candle should be bright enough to be detected at large redshifts. It should also have a luminosity that is well determined.[5]

[5] A useful cautionary tale in this regard is the saga of Edwin Hubble. In the 1929 paper that first demonstrated that $d_L \propto z$ when $z \ll 1$, Hubble underestimated the luminosity distances to galaxies by a factor of ~ 7 because he underestimated the luminosity of his standard candles by a factor of ~ 49.

One time-honored variety of standard candle is the class of *Cepheid variable stars*. Cepheids, as they are known, are highly luminous supergiant stars, with mean luminosities in the range $\bar{L} = 400 \rightarrow 40{,}000\,\mathrm{L}_\odot$. Cepheids are pulsationally unstable. As they pulsate radially, their luminosity varies in response, partially due to the change in their surface area, and partially due to the changes in the surface temperature as the star pulsates. The pulsational periods, as reflected in the observed brightness variations of the star, lie in the range $P = 1.5 \rightarrow 60$ days.

On the face of it, Cepheids don't seem sufficiently standardized to be standard candles; their mean luminosities range over two orders of magnitude. How can you tell whether you are looking at an intrinsically faint Cepheid ($L \approx 400\,\mathrm{L}_\odot$) or at an intrinsically bright Cepheid ($L \approx 40{,}000\,\mathrm{L}_\odot$) ten times farther away? The key to calibrating Cepheids was discovered by Henrietta Leavitt, at Harvard College Observatory. In the years prior to World War I, Leavitt was studying variable stars in the Large and Small Magellanic Clouds, a pair of relatively small satellite galaxies orbiting our own galaxy. For each Cepheid in the Small Magellanic Cloud (SMC), she measured the period P by finding the time between maxima in the observed brightness, and found the mean flux $\bar{f}$, averaged over one complete period. She noted that there was a clear relation between P and $\bar{f}$, with stars having the longest period of variability also having the largest flux. Since the depth of the SMC, front to back, is small compared to its distance from us, she was justified in assuming that the difference in mean flux for the Cepheids was due to differences in their mean luminosity, not differences in their luminosity distance. Leavitt had discovered a period–luminosity relation for Cepheid variable stars. If the same period–luminosity relation holds true for all Cepheids, in all galaxies, then Cepheids can act as a standard candle.

Suppose, for instance, you find a Cepheid star in the Large Magellanic Cloud (LMC) and another in M31. They both have a pulsational period of 10 days, so you assume, from the period–luminosity relation, that they have the same mean luminosity $\bar{L}$. By careful measurement, you determine that

$$\frac{\bar{f}_{\mathrm{LMC}}}{\bar{f}_{\mathrm{M31}}} = 230. \tag{7.43}$$

Thus, you conclude that the luminosity distance to M31 is greater than that to the LMC[6] by a factor

$$\frac{d_L(\mathrm{M31})}{d_L(\mathrm{LMC})} = \left(\frac{\bar{f}_{\mathrm{LMC}}}{\bar{f}_{\mathrm{M31}}}\right)^{1/2} = \sqrt{230} = 15.2. \tag{7.44}$$

Note that if you only know the relative fluxes of the two Cepheids, and not their luminosity $\bar{L}$, you will only know the *relative* distances of M31 and the LMC. To fix an absolute distance to M31, to the LMC, and to other galaxies containing

[6] In practice, given the intrinsic scatter in the period–luminosity relation, and the inevitable error in measuring fluxes, astronomers would not rely on a single Cepheid in each galaxy. Rather, they would measure $\bar{f}$ and P for as many Cepheids as possible in each galaxy, then find the ratio of luminosity distances that would make the period–luminosity relations for the two galaxies coincide.

Cepheids, you need to know the luminosity $\bar{L}$ for a Cepheid of a given period P. If, for instance, you could measure the parallax distance d_π to a Cepheid within our own galaxy, you could then compute its luminosity $\bar{L} = 4\pi d_\pi^2 \bar{f}$, and use it to normalize the period–luminosity relation for Cepheids.[7] Unfortunately, Cepheids are rare stars; only the very nearest Cepheids in our galaxy have had their distances measured with even modest accuracy by the Hipparcos satellite. The nearest Cepheid is Polaris, as it turns out, at $d_\pi = 130 \pm 10\,\text{pc}$. The next nearest is probably δ Cephei (the prototype after which all Cepheids are named), at $d_\pi = 300 \pm 50\,\text{pc}$. Future space-based astrometric observatories will allow parallax distances to be measured with an accuracy greater than that provided by Hipparcos. Until the distances (and hence the luminosities) of nearby Cepheids are known with this great accuracy, astronomers must still rely on alternate methods of normalizing the period–luminosity relation for Cepheids. The most usual method involves finding the distance to the Large Magellanic Cloud by secondary methods,[8] then using this distance to compute the mean luminosity of the LMC Cepheids. The current consensus is that the Large Magellanic Cloud has a luminosity distance $d_L = 50 \pm 3\,\text{kpc}$, implying a distance to M31 of $d_L = 760 \pm 50\,\text{kpc}$.

With the Hubble Space Telescope, the fluxes and periods of Cepheids can be accurately measured out to luminosity distances of $d_L \sim 20\,\text{Mpc}$. Observation of Cepheid stars in the Virgo cluster of galaxies, for instance, has yielded a distance $d_L(\text{Virgo}) = 300\,d_L(\text{LMC}) = 15\,\text{Mpc}$. One of the motivating reasons for building the Hubble Space Telescope in the first place was to use Cepheids to determine H_0. The net result of the Hubble Key Project to measure H_0 is displayed in Figure 2.5, showing that the Cepheid data are best fit with a Hubble constant of $H_0 = 75 \pm 8\,\text{km}\,\text{s}^{-1}\,\text{Mpc}^{-1}$.

There is a hidden difficulty involved in using Cepheid stars to determine H_0. Cepheids can take you out only to a distance $d_L \sim 20\,\text{Mpc}$; on this scale, the universe cannot be assumed to be homogeneous and isotropic. In fact, the Local Group is gravitationally attracted toward the Virgo cluster, causing it to have a peculiar motion in that direction. It is estimated, from dynamical models, that the recession velocity cz that we measure for the Virgo cluster is $250\,\text{km}\,\text{s}^{-1}$ less than it would be if the universe were perfectly homogeneous. The plot of cz versus d_L given in Figure 2.5 uses recession velocities that are corrected for this "Virgocentric flow," as it is called.

7.5 ■ STANDARD CANDLES AND THE ACCELERATING UNIVERSE

To determine the value of H_0 without having to worry about Virgocentric flow and other peculiar velocities, we need to determine the luminosity distance to standard candles with $d_L > 100\,\text{Mpc}$, or $z > 0.02$. To determine the value of

[7]Within our galaxy, which is not expanding, the parallax distance, the luminosity distance, and the proper distance are identical.

[8]A good review of these methods, and the distances they yield, is given by van den Bergh (2000).

q_0, we need to view standard candles for which the relation between d_L and z deviates significantly from the linear relation that holds true at lower redshifts. In terms of H_0 and q_0, the luminosity distance at small redshift is

$$d_L \approx \frac{c}{H_0} z \left[1 + \frac{1 - q_0}{2} z \right]. \tag{7.45}$$

At a redshift $z = 0.2$, for instance, the luminosity distance d_L in the Benchmark Model (with $q_0 = -0.55$) is 5% larger than d_L in an empty universe (with $q_0 = 0$).[9]

For a standard candle to be seen at $d_L > 100\,\text{Mpc}$ (to determine H_0 with minimal effects from peculiar velocity) or at $d_L > 1000\,\text{Mpc}$ (to determine q_0), it must be very luminous. Initial attempts to find a highly luminous standard candle focused on using entire galaxies as standard candles. This attempt foundered on the lack of standardization among galaxies. Not only do galaxies have a wide range of luminosities at the present moment, but any individual galaxy has a luminosity that evolves significantly with time. For instance, an isolated galaxy, after an initial outburst of star formation, will fade gradually with time, as its stars exhaust their nuclear fuel and become dim stellar remnants. A galaxy in a rich cluster, by contrast, can actually become more luminous with time, as it "cannibalizes" smaller galaxies by merging with them. For any particular galaxy, it's difficult to tell which effect dominates. Since the luminosity evolution of galaxies is imperfectly understood, they aren't particularly suitable for use as standard candles.

In recent years, the standard candle of choice among cosmologists has been *type Ia supernovae*. A supernova may be loosely defined as an exploding star. Early in the history of supernova studies, when little was known about their underlying physics, supernovae were divided into two classes, on the basis of their spectra. Type I supernovae contain no hydrogen absorption lines in their spectra; type II supernovae contain strong hydrogen absorption lines. Gradually, it was realized that all type II supernovae are the same species of beast; they are massive stars ($M > 8\,\text{M}_\odot$) whose cores collapse to form a black hole or neutron star when their nuclear fuel is exhausted. During the rapid collapse of the core, the outer layers of the star are thrown off into space. Type I supernovae are actually two separate species, called type Ia and type Ib. Type Ib supernovae, it is thought, are massive stars whose cores collapse after the hydrogen-rich outer layers of the star have been blown away in strong stellar winds. Thus, type Ib and type II supernovae are driven by very similar mechanisms—their differences are superficial, in the most literal sense. Type Ia supernovae, however, are something completely different. They occur in close binary systems where one of the two stars in the system is a white dwarf; that is, a stellar remnant that is supported against gravity by electron degeneracy pressure. The transfer of mass from the companion star to the white dwarf eventually nudges the white dwarf over the Chandrasekhar limit of $1.4\,\text{M}_\odot$; this is the maximum mass at which the electron degeneracy pressure can

[9] If you think, optimistically, that you can determine luminosity distances with an accuracy much better than 5%, then you won't have to go as deep into space to determine q_0 accurately.

support a white dwarf against its own self-gravity. When the Chandrasekhar limit is exceeded, the white dwarf starts to collapse until its increased density triggers a runaway nuclear fusion reaction. The entire white dwarf becomes a fusion bomb, blowing itself to smithereens; unlike type II supernovae, type Ia supernovae do not leave a condensed stellar remnant behind.

Within our galaxy, type Ia supernovae occur roughly once per century, on average. Although type Ia supernovae are not frequent occurrences locally, they are extraordinarily luminous, and hence can be seen to large distances. The luminosity of an average type Ia supernova, at peak brightness, is $L = 4 \times 10^9 \, L_\odot$; that's 100,000 times more luminous than even the brightest Cepheid. For a few days, a type Ia supernova in a moderately bright galaxy can outshine all the other stars in the galaxy combined. Since moderately bright galaxies can be seen at $z \sim 1$, this means that type Ia supernovae can also be seen at $z \sim 1$. Not only are type Ia supernovae bright standard candles, they are also reasonably standardized standard candles. Consider type Ia supernovae in the Virgo cluster. Although there's only one type Ia supernova per century in our own galaxy, the total luminosity of the Virgo cluster is a few hundred times that of our galaxy. Thus, every year you can expect a few type Ia supernovae to go off in the Virgo cluster. Several type Ia supernovae have been observed in the Virgo cluster in the recent past, and have been found to have similar fluxes at maximum brightness.

So far, type Ia supernovae sound like ideal standard candles; very luminous and very standardized. There's one complication, however. Observation of supernovae in galaxies whose distances have been well determined by Cepheids reveal that type Ia supernovae do not have identical luminosities. Instead of all having $L = 4 \times 10^9 \, L_\odot$, their peak luminosities lie in the fairly broad range $L \approx (3 \to 5) \times 10^9 \, L_\odot$. However, it has also been noted that the peak luminosity of a type Ia supernova is tightly correlated with the shape of its light curve. Type Ia supernovae with luminosities that shoot up rapidly and decline rapidly are less luminous than average at their peak; supernovae with luminosities that rise and fall in a more leisurely manner are more luminous than average. Thus, just as the period of a Cepheid tells you its luminosity, the rise and fall time of a type Ia supernova tells you its peak luminosity.

Recently, two research teams, the "Supernova Cosmology Project" and the "High-z Supernova Search Team," have been conducting searches for supernovae in distant galaxies. They have used the observed light curves and redshifts of type Ia supernovae to measure cosmological parameters. First, by observing type Ia supernovae at $z \sim 0.1$, the value of H_0 can be determined. The results of the different groups are in reasonable agreement with each other. If the distance to the Virgo cluster is pegged at $d_L = 15\,\text{Mpc}$, as indicated by the Cepheid results, then the observed supernovae fluxes and redshifts are consistent with $H_0 = 70 \pm 7\,\text{km}\,\text{s}^{-1}\,\text{Mpc}^{-1}$, the value of the Hubble constant that we have adopted in this text.

In addition, the supernova groups have been attempting to measure the acceleration (or deceleration) of the universe by observing type Ia supernovae at higher redshift. To present the most recent supernova results to you, I will have to introduce the "magnitude" system used by astronomers to express fluxes and luminosities. The magnitude system, like much else in astronomy, has its roots in ancient Greece. The Greek astronomer Hipparchus, in the second century BC, divided the stars into six classes, according to their apparent brightness. The brightest stars were of "first magnitude," the faintest stars visible to the naked eye were of "sixth magnitude," and intermediate stars were ranked as second, third, fourth, and fifth magnitude. Long after the time of Hipparchus, it was realized that the response of the human eye is roughly logarithmic, and that stars of the first magnitude have fluxes (at visible wavelengths) about 100 times greater than stars of the sixth magnitude. On the basis of this realization, the magnitude system was placed on a more rigorous mathematical basis.

Nowadays, the bolometric *apparent magnitude* of a light source is defined in terms of the source's bolometric flux as

$$m \equiv -2.5 \log_{10}(f/f_x), \tag{7.46}$$

where the reference flux f_x is set at the value $f_x = 2.53 \times 10^{-8}\,\text{watt}\,\text{m}^{-2}$. Thanks to the negative sign in the definition, a small value of m corresponds to a large flux f. For instance, the flux of sunlight at the Earth's location is $f = 1367\,\text{watts}\,\text{m}^{-2}$; the Sun thus has a bolometric apparent magnitude of $m = -26.8$. The choice of reference flux f_x constitutes a tip of the hat to Hipparchus, since for stars visible to the naked eye it typically yields $0 < m < 6$.

The bolometric *absolute magnitude* of a light source is defined as the apparent magnitude that it would have if it were at a luminosity distance of $d_L = 10\,\text{pc}$. Thus, a light source with luminosity L has a bolometric absolute magnitude

$$M \equiv -2.5 \log_{10}(L/L_x), \tag{7.47}$$

where the reference luminosity is $L_x = 78.7\,\text{L}_\odot$, since that is the luminosity of an object that produces a flux $f_x = 2.53 \times 10^{-8}\,\text{watt}\,\text{m}^{-2}$ when viewed from a distance of 10 parsecs. The bolometric absolute magnitude of the Sun is thus $M = 4.74$. Although the system of apparent and absolute magnitudes seems strange to the uninitiated, the apparent magnitude is really nothing more than a logarithmic measure of the flux, and the absolute magnitude is a logarithmic measure of the luminosity.

Given the definitions of apparent and absolute magnitude, the relation between an object's apparent magnitude and its absolute magnitude can be written in the form

$$M = m - 5 \log_{10}\left(\frac{d_L}{10\,\text{pc}}\right), \tag{7.48}$$

where d_L is the luminosity distance to the light source. If the luminosity distance is given in units of megaparsecs, this relation becomes

$$M = m - 5 \log_{10}\left(\frac{d_L}{1\,\text{Mpc}}\right) - 25. \tag{7.49}$$

Since astronomers frequently quote fluxes and luminosities in terms of apparent and absolute magnitudes, they find it convenient to quote luminosity distances in terms of the *distance modulus* to a light source. The distance modulus is defined as $m - M$, and is related to the luminosity distance by the relation

$$m - M = 5\log_{10}\left(\frac{d_L}{1\,\mathrm{Mpc}}\right) + 25. \tag{7.50}$$

The distance modulus of the Large Magellanic Cloud, for instance, at $d_L = 0.050\,\mathrm{Mpc}$, is $m - M = 18.5$. The distance modulus of the Virgo cluster, at $d_L = 15\,\mathrm{Mpc}$, is $m - M = 30.9$. When $z \ll 1$, the luminosity distance to a light source is

$$d_L \approx \frac{c}{H_0} z\left(1 + \frac{1 - q_0}{2} z\right). \tag{7.51}$$

Substituting this relation into equation (7.50), we have an equation that gives the relation between distance modulus and redshift:

$$m - M \approx 43.17 - 5\log_{10}\left(\frac{H_0}{70\,\mathrm{km\,s^{-1}\,Mpc^{-1}}}\right) + 5\log_{10} z + 1.086(1 - q_0)z. \tag{7.52}$$

For a population of standard candles with known luminosity L (and hence of known bolometric absolute magnitude M), we measure the flux f (or equivalently, the bolometric apparent magnitude m) and the redshift z. In the limit $z \rightarrow 0$, a plot of $m - M$ versus $\log_{10} z$ gives a straight line whose amplitude at a given value of z tells us the value of H_0. At slightly larger values of z, the deviation of the plot from a straight line tells us the value of q_0. At a given value of z, an accelerating universe (with $q_0 < 0$) yields standard candles with a smaller flux than would a decelerating universe (with $q_0 > 0$).

The upper panel of Figure 7.5 shows the plot of distance modulus versus redshift for the combined supernova samples of the High-z Supernova Search Team (given by the filled circles) and the Supernova Cosmology Project (given by the open circles). The observational results are compared to the expected results for three model universes. One universe is flat, and contains nothing but matter ($\Omega_{m,0} = 1$, $q_0 = 0.5$). The second is negatively curved, and contains nothing but matter ($\Omega_{m,0} = 0.3$, $q_0 = 0.15$). The third is flat, and contains both matter and a cosmological constant ($\Omega_{m,0} = 0.3$, $\Omega_{\Lambda,0} = 0.7$, $q_0 = -0.55$). The data are best fitted by the third of the models—which is, in fact, our Benchmark Model. The bottom panel of Figure 7.5 shows this result more clearly. It shows the difference between the data and the predictions of the negatively curved, matter-only model. The conclusion that the universe is accelerating derives from the observation that the supernovae seen at $z \sim 0.5$ are, on average, about 0.25 magnitudes fainter than they would be in a decelerating universe with $\Omega_{m,0} = 0.3$ and no cosmological constant.

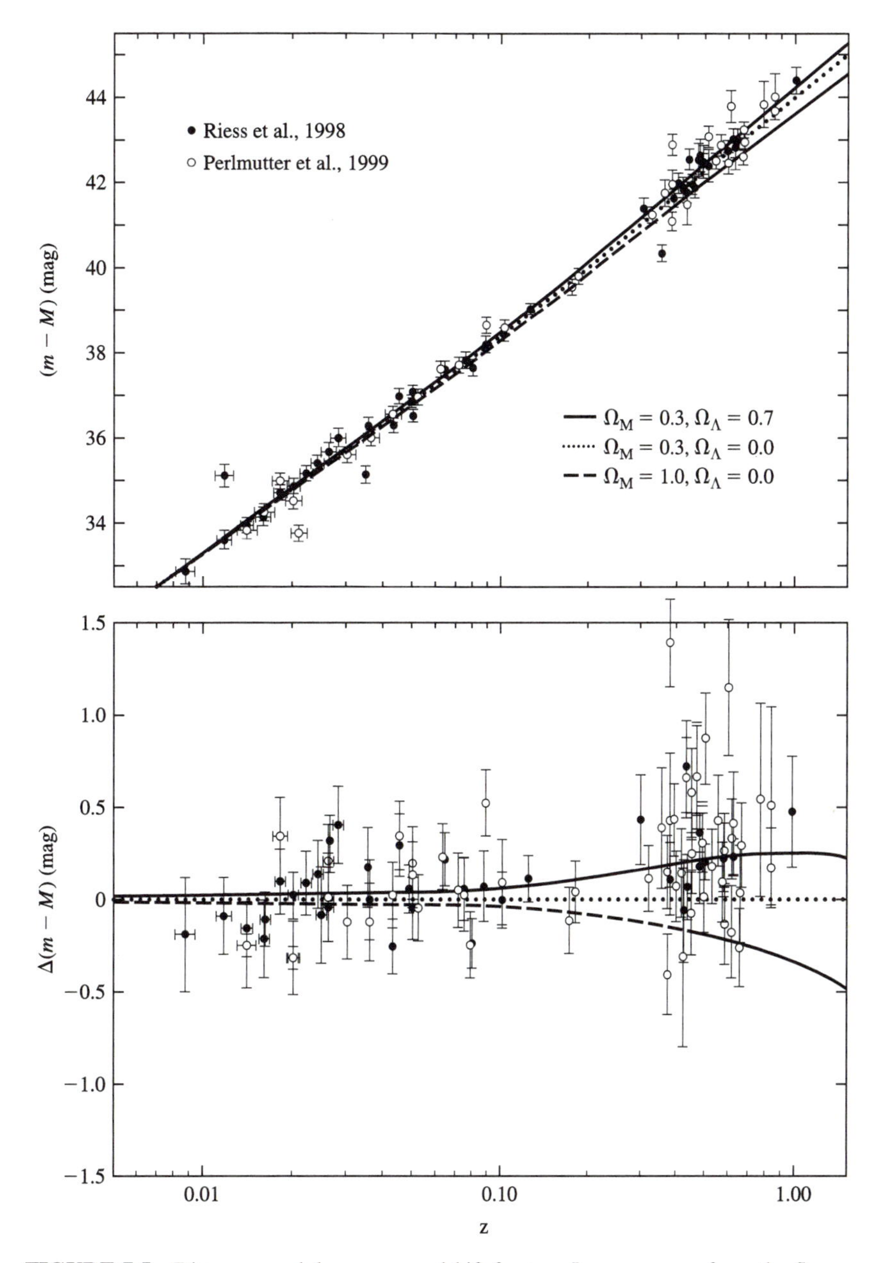

FIGURE 7.5 Distance modulus versus redshift for type Ia supernovae from the Supernova Cosmology Project (Perlmutter et al., 1999, ApJ, 517, 565) and the High-z Supernova Search Team (Riess et al., 1998, AJ, 116, 1009). The bottom panel shows the difference between the data and the predictions of a negatively curved $\Omega_{m,0} = 0.3$ model.

The supernova data extend out to $z \sim 1$; this is beyond the range where an expansion in terms of H_0 and q_0 is adequate to describe the scale factor $a(t)$. Thus, the two supernova teams customarily describe their results in terms of a model universe that contains both matter and a cosmological constant. After choosing values of $\Omega_{m,0}$ and $\Omega_{\Lambda,0}$, they compute the expected relation between $m - M$ and z, and compare it to the observed data. The results of fitting these model universes are given in Figure 7.6. The ovals drawn on Figure 7.6 enclose those values of $\Omega_{m,0}$ and $\Omega_{\Lambda,0}$ that give the best fit to the supernova data. The results of the two teams (the solid ovals and dotted ovals) give very similar results. Three concentric ovals are shown for each team's result; they correspond to 1σ, 2σ, and 3σ confidence intervals, with the inner oval representing the highest probability.

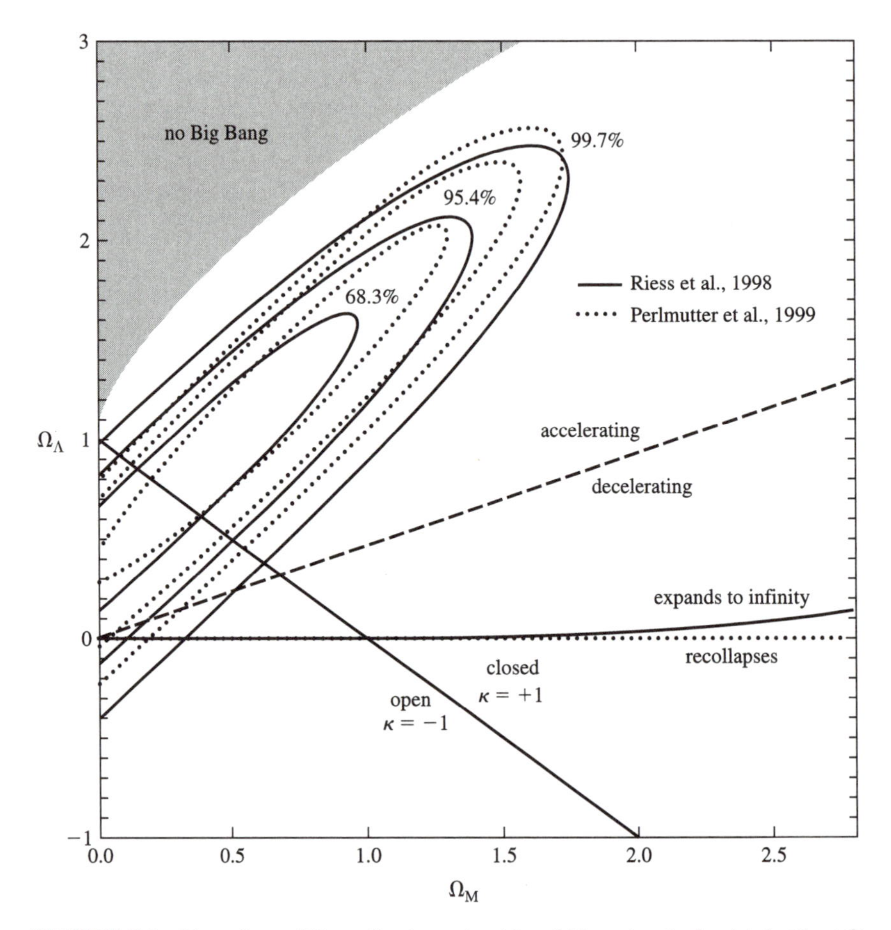

FIGURE 7.6 The values of $\Omega_{m,0}$ (horizontal axis) and $\Omega_{\Lambda,0}$ (vertical axis) that best fit the data shown in Figure 7.5. The solid ovals show the best-fitting values for the High-z Supernova Search Team data; the dotted ovals show the best-fitting values for the Supernova Cosmology Project data.

The best fitting models lie along the line $0.8\Omega_{m,0} - 0.6\Omega_{\Lambda,0} \approx -0.2$. Note that decelerating universes (with $q_0 > 0$) can be strongly excluded by the data, as can Big Crunch universes (labeled 'Recollapses' in Figure 7.6), and Big Bounce universes (labeled 'No Big Bang' in Figure 7.6). The supernova data are consistent with negative curvature (labeled 'Open' in Figure 7.6), positive curvature (labeled 'Closed' in Figure 7.6), or with a spatially flat universe.

The results of the supernova teams made headlines when they were first announced; the discovery of the accelerating universe was named by *Science* magazine as the 'Scientific Breakthrough of the Year' for 1998. It is prudent to remember, however, that all the hoopla about the accelerating universe is based on the observation that type Ia supernova at $z \sim 0.5$ and beyond have somewhat lower fluxes (by about 25%) than they would have in a decelerating universe. There are other reasons why their fluxes might be low. For instance, if type Ia supernovae were intrinsically less luminous at $z \sim 0.5$ than at $z \sim 0$, that could explain their low fluxes. (If a typical supernova at $z \sim 0.5$ had $L = 3 \times 10^9\,\mathrm{L}_\odot$ rather than $4 \times 10^9\,\mathrm{L}_\odot$, that would explain their observed dimness, without the need to invoke a cosmological constant. Conversely, if the typical supernova at $z \sim 0.5$ had $L = 5 \times 10^9\,\mathrm{L}_\odot$ rather than $4 \times 10^9\,\mathrm{L}_\odot$, that would require an even larger cosmological constant to explain their observed dimness.) However, the other properties of type Ia supernovae, such as their spectra, don't seem to evolve with time, so why should their luminosity? Perhaps the fluxes of supernovae at $z \sim 0.5$ are low because some of their light is scattered or absorbed by intervening dust. However, dust tends to scatter some wavelengths of light more than others. This would change the shape of the spectrum of distant type Ia supernovae, but no dependence of spectral shape on redshift is observed.

In sum, the supernova results of Figure 7.6 provide persuasive (but, given the caveats, not absolutely compelling) evidence for an accelerating universe. We will see in future chapters how additional observational evidence interlocks with the supernova results to suggest that we live in a nearly flat accelerating universe with $\Omega_{m,0} \approx 0.3$ and $\Omega_{\Lambda,0} \approx 0.7$.

SUGGESTED READING

Full references are given in the Annotated Bibliography on page 235.

Liddle (1999), ch. 6: The relation among H_0, q_0, Ω_0, and Λ.

Narlikar (2002), ch. 9, 10: Local observations ($z < 0.1$) and more distant observations ($z > 0.1$) of the universe, and what they tell us about cosmological parameters.

Peacock (1999), ch. 5: A review of distance measures used in cosmology.

Rich (2001), ch. 5.2: A summary of the supernova Ia results.

PROBLEMS

7.18. Suppose that a polar bear's foot has a luminosity of $L = 10$ watts. What is the bolometric absolute magnitude of the bear's foot? What is the bolometric apparent magnitude of the foot at a luminosity distance of $d_L = 0.5$ km? If a bolometer can detect the bear's foot at a maximum luminosity distance of $d_L = 0.5$ km, what is the maximum luminosity distance at which it could detect the Sun? What is the maximum luminosity distance at which it could detect a supernova with $L = 4 \times 10^9\,L_\odot$?

7.19. Suppose that a polar bear's foot has a diameter of $\ell = 0.16$ m. What is the angular size $\delta\theta$ of the foot at an angular-diameter distance of $d_A = 0.5$ km? In the Benchmark Model, what is the minimum possible angular size of the polar bear's foot?

7.20. Suppose that you are in a spatially flat universe containing a single component with a unique equation-of-state parameter w. What are the current proper distance $d_p(t_0)$, the luminosity distance d_L and the angular-diameter distance d_A as a function of z and w? At what redshift will d_A have a maximum value? What will this maximum value be, in units of the Hubble distance?

7.21. Verify that equation (7.52) is correct in the limit of small z. (You will probably want to use the relation $\log_{10}(1+x) \approx 0.4343 \ln(1+x) \approx 0.4343x$ in the limit $|x| \ll 1$.)

7.22. The surface brightness Σ of an astronomical object is defined as its observed flux divided by its observed angular area; thus, $\Sigma \propto f/(\delta\theta)^2$. For a class of objects that are both standard candles and standard yardsticks, what is Σ as a function of redshift? Would observing the surface brightness of this class of objects be a useful way of determining the value of the deceleration parameter q_0? Why or why not?

7.23. You observe a quasar at a redshift $z = 5.0$, and determine that the observed flux of light from the quasar varies on a timescale $\delta t_0 = 3$ days. If the observed variation in flux is due to a variation in the intrinsic luminosity of the quasar, what was the variation timescale δt_e at the time the light was emitted? For the light from the quasar to vary on a timescale δt_e, the bulk of the light must come from a region of physical size $R \leq R_{\text{max}} = c(\delta t_e)$. What is R_{max} for the observed quasar? What is the angular size of R_{max} in the Benchmark Model?

7.24. Derive the relation $A_p(t_0) = 4\pi S_\kappa(r)^2$, as given in equation (7.24), starting from the Robertson–Walker metric of equation (7.22).

7.25. A spatially flat universe contains a single component with equation-of-state parameter w. In this universe, standard candles of luminosity L are distributed homogeneously in space. The number density of the standard candles is n_0 at $t = t_0$, and the standard candles are neither created nor destroyed. Show that the observed flux from a single standard candle at redshift z is

$$f(z) = \frac{L(1+3w)^2}{16\pi(c/H_0)^2} \frac{1}{(1+z)^2} \left[1 - (1+z)^{-(1+3w)/2}\right]^{-2} \tag{7.53}$$

when $w \neq -\frac{1}{3}$. What is the corresponding relation when $w = -\frac{1}{3}$? Show that the observed intensity (that is, the power per unit area per steradian of sky) from standard candles with redshifts in the range $z \rightarrow z + dz$ is

$$dJ(z) = \frac{n_0 L(c/H_0)}{4\pi}(1+z)^{-(7+3w)/2} dz. \tag{7.54}$$

What will be the total intensity J of all standard candles integrated over all redshifts? Explain why the night sky is of finite brightness even in universes with $w \leq -\frac{1}{3}$, which have an infinite horizon distance.

CHAPTER 8

Dark Matter

Cosmologists, over the years, have dedicated a large amount of time and effort to determining the matter density of the universe. There are many reasons for this obsession. First, the density parameter in matter, $\Omega_{m,0}$, is important in determining the spatial curvature and expansion rate of the universe. Even if the cosmological constant is nonzero, the matter content of the universe is not negligible today, and was the dominant component in the fairly recent past. Another reason for wanting to know the matter density of the universe is to find out what the universe is made of. What fraction of the density is made of stars, and other familiar types of baryonic matter? What fraction of the density is made of dark matter? What constitutes the dark matter—cold stellar remnants, black holes, exotic elementary particles, or some other substance too dim for us to see? These questions, and others, have driven astronomers to take a census of the universe, find out what types of matter it contains, and in what quantities.

We have already seen in the previous chapter one method of putting limits on $\Omega_{m,0}$. The apparent magnitude (or flux) of type Ia supernovae as a function of redshift is consistent with a flat universe having $\Omega_{m,0} \approx 0.3$ and $\Omega_{\Lambda,0} \approx 0.7$. However, neither $\Omega_{m,0}$ nor $\Omega_{\Lambda,0}$ is individually well-constrained by the supernova observations. The supernova data are consistent with $\Omega_{m,0} = 0$ if $\Omega_{\Lambda,0} \approx 0.4$; they are also consistent with $\Omega_{m,0} = 1$ if $\Omega_{\Lambda,0} \approx 1.7$. In order to determine $\Omega_{m,0}$ more accurately, we will have to adopt alternate methods of estimating the matter content of the universe.

8.1 ■ VISIBLE MATTER

Some types of matter, such as stars, help astronomers to detect them by broadcasting photons in all directions. Stars emit light primarily in the infrared, visible, and ultraviolet range of the electromagnetic spectrum. Suppose, for instance, you install a B-band filter on your telescope. Such a filter allows only photons in the wavelength range $4.0 \times 10^{-7}\,\mathrm{m} < \lambda < 4.9 \times 10^{-7}\,\mathrm{m}$ to pass through.[1] The "B" in B-band stands for "blue"; however, in addition to admitting blue light, a B-

[1] For comparison, your eyes detect photons in the wavelength range $4 \times 10^{-7}\,\mathrm{m} < \lambda < 7 \times 10^{-7}\,\mathrm{m}$.

band filter also lets through violet light. The Sun's luminosity in the B band is $L_{\odot,B} = 4.7 \times 10^{25}$ watts.[2]

In the B band, the total luminosity density of stars within a few hundred megaparsecs of our galaxy is

$$j_{\star,B} = 1.2 \times 10^{8}\,L_{\odot,B}\,\text{Mpc}^{-3}. \tag{8.1}$$

To convert a luminosity density $j_{\star,B}$ into a mass density $\rho_\star$, we need to know the *mass-to-light ratio* for the stars. That is, we need to know how many kilograms of star, on average, it takes to produce one watt of starlight in the B band. If all stars were identical to the Sun, we could simply say that there is one solar mass of stars for each solar luminosity of output power, or $\langle M/L_B\rangle = 1\,M_\odot/L_{\odot,B}$. However, stars are not uniform in their properties. They have a wide range of masses and a wider range of B-band luminosities. For main sequence stars, powered by hydrogen fusion in their centers, the mass-to-light ratio ranges from $M/L_B \sim 10^{-3}\,M_\odot/L_{\odot,B}$ for the brightest, most massive stars (the O stars in the classic OBAFGKM spectral sequence) to $M/L_B \sim 10^{3}\,M_\odot/L_{\odot,B}$ for the dimmest, least massive stars (the M stars).

Thus, the mass-to-light ratio of the stars in a galaxy will depend on the mix of stars that it contains. As a first guess, let's suppose that the mix of stars in the solar neighborhood is not abnormal. Within 1 kiloparsec of the Sun, the mass-to-light ratio of the stars works out to be

$$\langle M/L_B\rangle \approx 4\,M_\odot/L_{\odot,B} \approx 170{,}000\,\text{kg watt}^{-1}. \tag{8.2}$$

Although a mass-to-light ratio of 170 tons per watt doesn't seem, at first glance, like a very high efficiency, you must remember that the mass of a star includes all the fuel that it will require during its entire lifetime.

If the mass-to-light ratio of the stars within a kiloparsec of us is not unusually high or low, then the mass density of stars in the universe is

$$\rho_{\star,0} = \langle M/L_B\rangle j_{\star,B} \approx 5 \times 10^{8}\,M_\odot\,\text{Mpc}^{-3}. \tag{8.3}$$

Since the current critical density of the universe is equivalent to a mass density of $\rho_{c,0} = \varepsilon_{c,0}/c^2 = 1.4 \times 10^{11}\,M_\odot\,\text{Mpc}^{-3}$, the current density parameter of stars is

$$\Omega_{\star,0} = \frac{\rho_{\star,0}}{\rho_{c,0}} \approx \frac{5 \times 10^{8}\,M_\odot\,\text{Mpc}^{-3}}{1.4 \times 10^{11}\,M_\odot\,\text{Mpc}^{-3}} \approx 0.004. \tag{8.4}$$

Stars make up less than $\frac{1}{2}$% of the density necessary to flatten the universe. In truth, the number $\Omega_{\star,0} \approx 0.004$ is not a precisely determined one, largely because of the uncertainty in the number of low-mass, low-luminosity stars in galaxies. In our galaxy, for instance, $\sim 95\%$ of the stellar luminosity comes from stars more luminous than the Sun, but $\sim 80\%$ of the stellar mass comes from stars *less* lumi-

[2]This is only 12% of the Sun's total luminosity. About 6% of the luminosity is emitted at ultraviolet wavelengths, and the remaining 82% is emitted at wavelengths too long to pass through the B filter.

nous than the Sun. The density parameter in stars will be further increased if you include in the category of "stars" stellar remnants (such as white dwarfs, neutron stars, and black holes) and brown dwarfs. A brown dwarf is a self-gravitating ball of gas, too low in mass to sustain nuclear fusion in its interior. Because brown dwarfs and isolated cool stellar remnants are difficult to detect, their number density is not well determined.

Galaxies also contain baryonic matter that is not in the form of stars, stellar remnants, or brown dwarfs. The interstellar medium contains significant amounts of gas. In our galaxy and in M31, for instance, the mass of interstellar gas is roughly equal to 10% of the mass of stars. In irregular galaxies such as the Magellanic Clouds, the ratio of gas to stars is even higher. In addition, there is a significant amount of gas between galaxies. Consider a rich cluster of galaxies such as the Coma cluster, located 100 Mpc from our galaxy, in the direction of the constellation Coma Berenices. At visible wavelengths, as shown in Figure 8.1, most of the light comes from the stars in the cluster's galaxies. The Coma cluster contains thousands of galaxies; their summed luminosity in the B band comes to $L_{\text{Coma},B} = 8 \times 10^{12}\,\text{L}_{\odot,B}$. If the mass-to-light ratio of the stars in the Coma cluster is $\langle M/L_B \rangle \approx 4\,\text{M}_\odot/\text{L}_{\odot,B}$, then the total mass of stars in the Coma cluster is $M_{\text{Coma},\star} \approx 3 \times 10^{13}\,\text{M}_\odot$. Although 30 trillion solar masses represents a lot of stars, the stellar mass in the Coma cluster is small compared to the mass of the hot, intracluster gas between the galaxies in the cluster. X-ray images, such as the

FIGURE 8.1 The Coma cluster as seen in visible light. The image shown is 35 arcminutes across, equivalent to $\sim$ 1 Mpc at the distance of the Coma cluster. [From the Space Telescope Science Institute Digitized Sky Survey, ©AURA/STScI]

FIGURE 8.2 The Coma cluster as seen in x-ray light. The scale is the same as that of the previous image. [From the ROSAT x-ray observatory; courtesy Max-Planck-Institut für extraterrestriche Physik. This figure and the previous figure were produced by Raymond White, using NASA's SkyView facility.]

one shown in Figure 8.2, reveal that hot, low-density gas, with a typical temperature of $T \approx 1 \times 10^8$ K, fills the space between clusters, emitting x-rays with a typical energy of $E \sim kT_{\rm gas} \sim 9\,{\rm keV}$. The total amount of x-ray emitting gas in the Coma cluster is estimated to be $M_{\rm Coma,gas} \approx 2 \times 10^{14}\,{\rm M}_\odot$, roughly six or seven times the mass in stars.

As it turns out, the best current limits on the baryon density of the universe come from the predictions of primordial nucleosynthesis. As we will see in Chapter 10, the efficiency with which fusion takes place in the early universe, converting hydrogen into deuterium, helium, lithium, and other elements, depends on the density of protons and neutrons present. Detailed studies of the amounts of deuterium and other elements present in primordial gas clouds indicate that the density parameter of baryonic matter must be

$$\Omega_{\rm bary,0} = 0.04 \pm 0.01, \tag{8.5}$$

an order of magnitude larger than the density parameter for stars. When you stare up at the night sky and marvel at the glory of the stars, you are actually marveling at a minority of the baryonic matter in the universe. Most of the baryons are too cold to be readily visible (the infrared emitting brown dwarfs and cold stellar remnants) or too diffuse to be readily visible (the low density x-ray gas in clusters).

8.2 ■ DARK MATTER IN GALAXIES

The situation, in fact, is even more extreme than stated in the previous section. Not only is most of the baryonic matter undetectable by our eyes, but most of the matter is not even baryonic. The majority of the matter in the universe is *nonbaryonic dark matter*, which doesn't absorb, emit, or scatter light of any wavelength. One way of detecting dark matter is to look for its gravitational influence on visible matter. A classic method of detecting dark matter involves looking at the orbital speeds of stars in spiral galaxies such as our own galaxy and M31. Spiral galaxies contain flattened disks of stars; within the disk, stars are on nearly circular orbits around the center of the galaxy. The Sun, for instance, is on such an orbit—it is $R = 8.5\,\text{kpc}$ from the galactic center, and has an orbital speed of $v = 220\,\text{km}\,\text{s}^{-1}$.

Suppose that a star is on a circular orbit around the center of its galaxy. If the radius of the orbit is R and the orbital speed is v, then the star experiences an acceleration

$$a = \frac{v^2}{R}, \tag{8.6}$$

directed toward the center of the galaxy. If the acceleration is provided by the gravitational attraction of the galaxy, then

$$a = \frac{GM(R)}{R^2}, \tag{8.7}$$

where $M(R)$ is the mass contained within a sphere of radius R centered on the galactic center.[3] The relation between v and M is found by setting equation (8.6) equal to equation (8.7):

$$\frac{v^2}{R} = \frac{GM(R)}{R^2}, \tag{8.8}$$

or

$$v = \sqrt{\frac{GM(R)}{R}}. \tag{8.9}$$

The surface brightness I of the disk of a spiral galaxy typically falls off exponentially with distance from the center:

$$I(R) = I(0)\exp\left(-\frac{R}{R_s}\right), \tag{8.10}$$

with the scale length R_s typically being a few kiloparsecs. For our galaxy, $R_s \approx 4\,\text{kpc}$; for M31, a somewhat larger disk galaxy, $R_s \approx 6\,\text{kpc}$. Once you are a

[3]Equation (8.7) assumes that the mass distribution of the galaxy is spherically symmetric. This is not, strictly speaking, true (the stars in the disk obviously have a flattened distribution), but the flattening of the galaxy provides only a small correction to the equation for the gravitational acceleration.

FIGURE 8.3 An observer sees a disk at an inclination angle i.

few scale lengths from the center of the spiral galaxy, the mass of stars inside R becomes essentially constant. Thus, if stars contributed all, or most, of the mass in a galaxy, the velocity would fall as $v \propto 1/\sqrt{R}$ at large radii. This relation between orbital speed and orbital radius, $v \propto 1/\sqrt{R}$, is referred to as "Keplerian rotation," since it's what Kepler found for orbits in the solar system, where the mass is strongly concentrated toward the center.[4]

The orbital speed v of stars within a spiral galaxy can be determined from observations. Consider a galaxy with the shape of a thin circular disk. In general, we won't be seeing the disk perfectly face-on or edge-on; we'll see it at an inclination i, where i is the angle between our line of sight to the disk and a line perpendicular to the disk (see Figure 8.3). The disk we see in projection will be elliptical, not circular, with an axis ratio

$$b/a = \cos i. \tag{8.11}$$

For example, the galaxy M31 looks extremely elongated as seen from Earth, with an observed axis ratio $b/a = 0.22$. This indicates that we are seeing M31 fairly close to edge-on, with an inclination $i = \cos^{-1}(0.22) = 77°$. By measuring the redshift of the absorption, or emission, lines in light from the disk, we can find the radial velocity $v_r(R) = cz(R)$ along the apparent long axis of the galaxy. Since the redshift contains only the component of the stars' orbital velocity that lies along the line of sight, the radial velocity we measure will be

$$v_r(R) = v_{\text{gal}} + v(R)\sin i, \tag{8.12}$$

[4]99.8% of the solar system's mass is contained within the Sun.

where v_{gal} is the radial velocity of the galaxy as a whole, resulting from the expansion of the universe, and $v(R)$ is the orbital speed at a distance R from the center of the disk. We can thus compute the orbital speed $v(R)$ in terms of observable properties as

$$v(R) = \frac{v_r(R) - v_{gal}}{\sin i} = \frac{v_r(R) - v_{gal}}{\sqrt{1 - b^2/a^2}}. \tag{8.13}$$

The first astronomer to detect the rotation of M31 was Vesto Slipher, in 1914. However, given the difficulty of measuring the spectra at low surface brightness, the orbital speed v at $R > 3R_s = 18\,\text{kpc}$ was not accurately measured until more than half a century later. In 1970, Vera Rubin and Kent Ford looked at emission lines from regions of hot ionized gas in M31, and were able to find the orbital speed $v(R)$ out to a radius $R = 24\,\text{kpc} = 4R_s$. Their results, shown as the open circles in Figure 8.4, give no sign of a Keplerian decrease in the orbital speed. Beyond $R = 4R_s$, the visible light from M31 was too faint for Rubin and Ford to measure the redshift; as they wrote in their original paper, "extrapolation beyond that distance is a matter of taste." At $R > 4R_s$, a small amount of atomic hydrogen is still in the disk of M31, which can be detected by means of its emission line at $\lambda = 21\,\text{cm}$. By measuring the redshift of this emission line, M. Roberts and R. Whitehurst found that the orbital speed stayed at a nearly constant value of $v(R) \approx 230\,\text{km}\,\text{s}^{-1}$ out to $R \approx 30\,\text{kpc} \approx 5R_s$, as shown by the solid dots in Figure 8.4. Since the orbital speed of the stars and gas at large radii ($R > 3R_s$) is greater than it would be if stars and gas were the only matter present, we deduce the presence of a *dark halo* within which the visible stellar disk is embedded. The

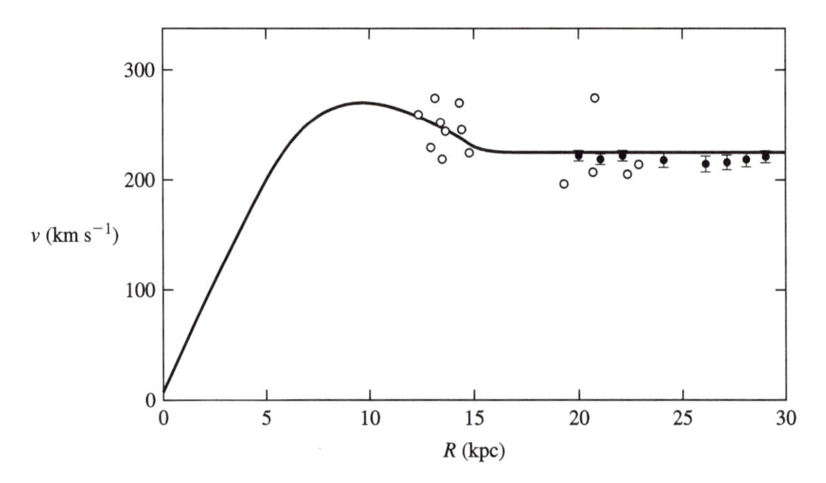

FIGURE 8.4 The orbital speed v as a function of radius in M31. The open circles show the results of Rubin and Ford (1970, ApJ, 159, 379) at visible wavelengths; the solid dots with error bars show the results of Roberts and Whitehurst (1975, ApJ, 201, 327) at radio wavelengths.

mass of the dark halo provides the necessary gravitational "anchor" to keep the high-speed stars and gas from being flung out into intergalactic space.

M31 is not a freak; most, if not all, spiral galaxies have comparable dark halos. For instance, our own galaxy has an orbital speed that actually seems to be rising slightly at $R > 15\,\text{kpc}$, instead of decreasing in a Keplerian fashion. Thousands of spiral galaxies have had their orbital velocities $v(R)$ measured; typically, v is roughly constant at $R > R_s$. If we approximate the orbital speed v as being constant with radius, the mass of a spiral galaxy, including both the luminous disk and the dark halo, can be found from equation (8.9):

$$M(R) = \frac{v^2 R}{G} = 9.6 \times 10^{10}\,\text{M}_\odot \left(\frac{v}{220\,\text{km s}^{-1}}\right)^2 \left(\frac{R}{8.5\,\text{kpc}}\right). \tag{8.14}$$

The values of v and R in the above equation are scaled to the Sun's location in our galaxy. Since our galaxy's luminosity in the B band is estimated to be $L_{\text{Gal},B} = 2.3 \times 10^{10}\,\text{L}_{\odot,B}$, this means that the mass-to-light ratio of our galaxy, taken as a whole, is

$$\langle M/L_B \rangle_{\text{Gal}} \approx 50\,\text{M}_\odot/\,\text{L}_{\odot,B} \left(\frac{R_{\text{halo}}}{100\,\text{kpc}}\right), \tag{8.15}$$

using $v = 220\,\text{km s}^{-1}$ in equation (8.14). The quantity R_{halo} is the radius of the dark halo surrounding the luminous disk of our galaxy. The exact value of R_{halo} is poorly known. At $R \approx 20\,\text{kpc}$, where the last detectable gas exists in the disk of our galaxy, the orbital speed shows no sign of a Keplerian decrease; thus, $R_{\text{halo}} > 20\,\text{kpc}$. A rough estimate of the halo size can be made by looking at the velocities of the globular clusters and satellite galaxies (such as the Magellanic Clouds) that orbit our galaxy. For these hangers-on to remain gravitationally bound to our galaxy, the halo must extend as far as $R_{\text{halo}} \approx 75\,\text{kpc}$, implying a total mass for our galaxy of $M_{\text{Gal}} \approx 8 \times 10^{11}\,\text{M}_\odot$, and a total mass-to-light ratio $\langle M/L_B \rangle_{\text{Gal}} \approx 40\,\text{M}_\odot/\,\text{L}_{\odot,B}$. This mass-to-light ratio is ten times greater than that of the stars in our galaxy, implying that the dark halo is an order of magnitude more massive than the stellar disk. Some astronomers have speculated that the dark halo is actually four times larger in radius, with $R_{\text{halo}} \approx 300\,\text{kpc}$; this would mean that our halo stretches nearly halfway to M31. With $R_{\text{halo}} \approx 300\,\text{kpc}$, the mass of our galaxy would be $M_{\text{Gal}} \approx 3 \times 10^{12}\,\text{M}_\odot$, and the total mass-to-light ratio would be $\langle M/L_B \rangle_{\text{Gal}} \approx 150\,\text{M}_\odot/\,\text{L}_{\odot,B}$.

If our galaxy is typical in having a dark halo 10 to 40 times more massive than its stellar component, then the density parameter of galaxies (including their dark halos) must be

$$\Omega_{\text{gal},0} = (10 \rightarrow 40)\Omega_{\star,0} \approx 0.04 \rightarrow 0.16. \tag{8.16}$$

Although the total density of galaxies is poorly known, given the uncertainty in the extent of their dark halos, it is likely to be larger than the density of baryons, $\Omega_{\text{bary},0} = 0.04 \pm 0.01$. Thus, some part of the dark halos of galaxies is likely to be comprised of *nonbaryonic* dark matter.

8.3 ■ DARK MATTER IN CLUSTERS

The first astronomer to make a compelling case for the existence of large quantities of dark matter was Fritz Zwicky, in the 1930's. In studying the Coma cluster of galaxies (shown in Figure 8.1), he noted that the dispersion in the radial velocity of the cluster's galaxies was very large—around $1000\,\mathrm{km\,s^{-1}}$. The stars and gas visible within the galaxies simply did not provide enough gravitational attraction to hold the cluster together. In order to keep the galaxies in the Coma cluster from flying off into the surrounding voids, Zwicky concluded, the cluster must contain a large amount of "dunkle Materie," or (translated into English) "dark matter."[5]

To follow Zwicky's reasoning at a more mathematical level, let us suppose that a cluster of galaxies is comprised of N galaxies, each of which can be approximated as a point mass, with a mass m_i $(i = 1, 2, \ldots, N)$, a position $\vec{x}_i$, and a velocity $\dot{\vec{x}}_i$. Clusters of galaxies are gravitationally bound objects, not expanding with the Hubble flow. They are small compared to the horizon size; the radius of the Coma cluster is $R_{\mathrm{Coma}} \approx 3\,\mathrm{Mpc} \approx 0.0002 d_{\mathrm{hor}}$. The galaxies within a cluster are moving at nonrelativistic speeds; the velocity dispersion within the Coma cluster is $\sigma_{\mathrm{Coma}} \approx 900\,\mathrm{km\,s^{-1}} \approx 0.003c$. Because of these considerations, we can treat the dynamics of the Coma cluster, and other clusters of galaxies, in a Newtonian manner. The acceleration of the ith galaxy in the cluster, then, is given by the Newtonian formula

$$\ddot{\vec{x}}_i = G \sum_{j \neq i} m_j \frac{\vec{x}_j - \vec{x}_i}{|\vec{x}_j - \vec{x}_i|^3}. \tag{8.17}$$

Note that equation (8.17) assumes that the cluster is an isolated system, with the gravitational acceleration due to matter outside the cluster being negligibly small.

The gravitational *potential energy* of the system of N galaxies is

$$W = -\frac{G}{2} \sum_{\substack{i,j \\ j \neq i}} \frac{m_i m_j}{|\vec{x}_j - \vec{x}_i|}. \tag{8.18}$$

This is the energy that would be required to pull the N galaxies away from each other so that they would all be at infinite distance from each other. (The factor of $\frac{1}{2}$ in front of the double summation ensures that each pair of galaxies is only counted once in computing the potential energy.) The potential energy of the cluster can also be written in the form

$$W = -\alpha \frac{GM^2}{r_h}, \tag{8.19}$$

[5]Although Zwicky's work popularized the phrase "dark matter," he was not the first to use it in an astronomical context. For instance, in 1908, Henri Poincaré discussed the possible existence within our galaxy of "matière obscure" (translated as "dark matter" in the standard edition of Poincaré's works).

where $M = \sum m_i$ is the total mass of all the galaxies in the cluster, α is a numerical factor of order unity that depends on the density profile of the cluster, and r_h is the *half-mass* radius of the cluster—that is, the radius of a sphere centered on the cluster's center of mass and containing a mass $M/2$. For observed clusters of galaxies, it is found that $\alpha \approx 0.4$ gives a good fit to the potential energy.

The *kinetic energy* associated with the relative motion of the galaxies in the cluster is

$$K = \frac{1}{2} \sum_i m_i |\dot{\vec{x}}_i|^2. \tag{8.20}$$

The kinetic energy K can also be written in the form

$$K = \frac{1}{2} M \langle v^2 \rangle, \tag{8.21}$$

where

$$\langle v^2 \rangle \equiv \frac{1}{M} \sum_i m_i |\dot{\vec{x}}_i|^2 \tag{8.22}$$

is the mean square velocity (weighted by galaxy mass) of all the galaxies in the cluster.

It is also useful to define the *moment of inertia* of the cluster as

$$I \equiv \sum_i m_i |\vec{x}_i|^2. \tag{8.23}$$

The moment of inertia I can be linked to the kinetic energy and the potential energy if we start by taking the second time derivative of I:

$$\ddot{I} = 2 \sum_i m_i (\vec{x}_i \cdot \ddot{\vec{x}}_i + \dot{\vec{x}}_i \cdot \dot{\vec{x}}_i). \tag{8.24}$$

Using equation (8.20), we can rewrite this as

$$\ddot{I} = 2 \sum_i m_i (\vec{x}_i \cdot \ddot{\vec{x}}_i) + 4K. \tag{8.25}$$

To introduce the potential energy W into the above relation, we can use equation (8.17) to write

$$\sum_i m_i (\vec{x}_i \cdot \ddot{\vec{x}}_i) = G \sum_{\substack{i,j \\ j \neq i}} m_i m_j \frac{\vec{x}_i \cdot (\vec{x}_j - \vec{x}_i)}{|\vec{x}_j - \vec{x}_i|^3}. \tag{8.26}$$

However, we could equally well switch around the i and j subscripts to find the equally valid equation

$$\sum_j m_j(\vec{x}_j \cdot \ddot{\vec{x}}_j) = G \sum_{\substack{j,i \\ i \neq j}} m_j m_i \frac{\vec{x}_j \cdot (\vec{x}_i - \vec{x}_j)}{|\vec{x}_i - \vec{x}_j|^3}. \tag{8.27}$$

Since

$$\sum_i m_i(\vec{x}_i \cdot \ddot{\vec{x}}_i) = \sum_j m_j(\vec{x}_j \cdot \ddot{\vec{x}}_j) \tag{8.28}$$

(it doesn't matter whether we call the variable over which we're summing i or j or k or "Fred"), we can combine equations (8.26) and (8.27) to find

$$\sum_i m_i(\vec{x}_i \cdot \ddot{\vec{x}}_i) = \frac{1}{2}\left[\sum_i m_i(\vec{x}_i \cdot \ddot{\vec{x}}_i) + \sum_j m_j(\vec{x}_j \cdot \ddot{\vec{x}}_j)\right]$$

$$= -\frac{G}{2} \sum_{\substack{i,j \\ j \neq i}} \frac{m_i m_j}{|\vec{x}_j - \vec{x}_i|} = W. \tag{8.29}$$

Thus, the first term on the right-hand side of equation (8.25) is simply $2W$, and we may now write down the simple relation

$$\ddot{I} = 2W + 4K. \tag{8.30}$$

This relation is known as the *virial theorem*. It was actually first derived in the nineteenth century in the context of the kinetic theory of gases, but as we have seen, it applies perfectly well to a self-gravitating system of point masses.

The virial theorem is particularly useful when it is applied to a system in steady state, with a constant moment of inertia. (This implies, among other things, that the system is neither expanding nor contracting, and that we are using a coordinate system in which the center of mass of the cluster is at rest.) If $I =$ constant, then the *steady-state virial theorem* is

$$0 = W + 2K, \tag{8.31}$$

or

$$K = -\frac{W}{2}. \tag{8.32}$$

That is, for a self-gravitating system in steady state, the kinetic energy K is equal to $-\frac{1}{2}$ times the potential energy W. Using equation (8.19) and (8.21) in equation (8.32), we find

$$\frac{1}{2}M\langle v^2\rangle = \frac{\alpha}{2}\frac{GM^2}{r_h}. \tag{8.33}$$

This means we can use the virial theorem to estimate the mass of a cluster of galaxies, or any other self-gravitating steady-state system:

$$M = \frac{\langle v^2 \rangle r_h}{\alpha G}. \tag{8.34}$$

Note the similarity between equation (8.14), used to estimate the mass of a rotating spiral galaxy, and equation (8.34), used to estimate the mass of a cluster of galaxies. In either case, we estimate the mass of a self-gravitating system by multiplying the square of a characteristic velocity by a characteristic radius, then dividing by the gravitational constant G.

Applying the virial theorem to a real cluster of galaxies, such as the Coma cluster, is complicated by the fact that we have only partial information about the cluster, and thus do not know $\langle v^2 \rangle$ and r_h exactly. For instance, we can find the line-of-sight velocity of each galaxy from its redshift, but the velocity perpendicular to the line of sight is unknown. From measurements of the redshifts of hundreds of galaxies in the Coma cluster, the mean redshift of the cluster is found to be

$$\langle z \rangle = 0.0232, \tag{8.35}$$

which can be translated into a radial velocity

$$\langle v_r \rangle = c\langle z \rangle = 6960\,\text{km}\,\text{s}^{-1} \tag{8.36}$$

and a distance

$$d_{\text{Coma}} = (c/H_0)\langle z \rangle = 99\,\text{Mpc}. \tag{8.37}$$

The velocity dispersion of the cluster along the line of sight is found to be

$$\sigma_r = \langle (v_r - \langle v_r \rangle)^2 \rangle^{1/2} = 880\,\text{km}\,\text{s}^{-1}. \tag{8.38}$$

If we assume that the velocity dispersion is isotropic, then the three-dimensional mean square velocity $\langle v^2 \rangle$ will be equal to three times the one-dimensional mean square velocity σ_r^2, yielding

$$\langle v^2 \rangle = 3(880\,\text{km}\,\text{s}^{-1})^2 = 2.32 \times 10^{12}\,\text{m}^2\,\text{s}^{-2}. \tag{8.39}$$

Estimating the half-mass radius r_h of the Coma cluster is even more peril-ridden than estimating the mean square velocity $\langle v^2 \rangle$. After all, we don't know the distribution of dark matter in the cluster beforehand; in fact, the total amount of dark matter is what we're trying to find out. However, if we assume that the mass-to-light ratio is constant with radius, then the sphere containing half the mass of the cluster will be the same as the sphere containing half the luminosity of the cluster. If we further assume that the cluster is intrinsically spherical, then the observed distribution of galaxies within the Coma cluster indicates a half-mass radius

$$r_h \approx 1.5\,\text{Mpc} \approx 4.6 \times 10^{22}\,\text{m}. \tag{8.40}$$

After all these assumptions and approximations, we may estimate the mass of the Coma cluster to be

$$M_{\text{Coma}} = \frac{\langle v^2 \rangle r_h}{\alpha G} \approx \frac{(2.32 \times 10^{12}\,\text{m}^2\,\text{s}^{-2})(4.6 \times 10^{22}\,\text{m})}{(0.4)(6.7 \times 10^{-11}\,\text{m}^3\,\text{s}^{-2}\,\text{kg}^{-1})} \tag{8.41}$$

$$\approx 4 \times 10^{45}\,\text{kg} \approx 2 \times 10^{15}\,\text{M}_\odot. \tag{8.42}$$

Thus, less than two percent of the mass of the Coma cluster consists of stars ($M_{\text{Coma},\star} \approx 3 \times 10^{13}\,\text{M}_\odot$), and only ten percent consists of hot intracluster gas ($M_{\text{Coma,gas}} \approx 2 \times 10^{14}\,\text{M}_\odot$). Combined with the luminosity of the Coma cluster, $L_{\text{Coma},B} = 8 \times 10^{12}\,\text{L}_{\odot,B}$, the total mass of the Coma cluster implies a mass-to-light ratio of

$$\left\langle \frac{M}{L_B} \right\rangle_{\text{Coma}} \approx \frac{250\,\text{M}_\odot}{\text{L}_{\odot,B}}, \tag{8.43}$$

greater than the mass-to-light ratio of our galaxy.

The presence of a vast reservoir of dark matter in the Coma cluster is confirmed by the fact that the hot, x-ray emitting intracluster gas, shown in Figure 8.2, is still in place; if there were no dark matter to anchor the gas gravitationally, the hot gas would have expanded beyond the cluster on time scales much shorter than the Hubble time. The temperature and density of the hot gas in the Coma cluster can be used to make yet another estimate of the cluster's mass. If the hot intracluster gas is supported by its own pressure against gravitational infall, it must obey the equation of hydrostatic equilibrium:

$$\frac{dP}{dr} = -\frac{GM(r)\rho(r)}{r^2}, \tag{8.44}$$

where P is the pressure of the gas, ρ is the density of the gas, and M is the *total* mass inside a sphere of radius r, including gas, stars, dark matter, lost socks, and anything else.[6] Of course, the gas in Coma isn't perfectly spherical in shape, as equation (8.44) assumes, but it's close enough to spherical to give a reasonable approximation to the mass.

The pressure of the gas is given by the perfect gas law,

$$P = \frac{\rho k T}{\mu m_p}, \tag{8.45}$$

where T is the temperature of the gas, and μ is its mass in units of the proton mass (m_p). The mass of the cluster, as a function of radius, is found by combining equations (8.44) and (8.45):

$$M(r) = \frac{kT(r)r}{G\mu m_p}\left[-\frac{d\ln\rho}{d\ln r} - \frac{d\ln T}{d\ln r}\right]. \tag{8.46}$$

[6] Equation (8.44) is the same equation that determines the internal structure of a star, where the inward force due to gravity is also exactly balanced by an outward force due to a pressure gradient.

The above equation assumes that μ is constant with radius, as we'd expect if the chemical composition and ionization state of the gas is uniform throughout the cluster.

The x-rays emitted from the hot intracluster gas are a combination of bremsstrahlung emission (caused by the acceleration of free electrons by protons and helium nuclei) and line emission from highly ionized iron and other heavy elements. Starting from an x-ray spectrum, it is possible to fit models to the emission and thus compute the temperature, density, and chemical composition of the gas. In the Coma cluster, for instance, temperature maps reveal relatively cool regions (at $kT \approx 5\,\text{keV}$) as well as hotter regions (at $kT \approx 12\,\text{keV}$), averaging to $kT \approx 9\,\text{keV}$ over the entire cluster. The mass of the Coma cluster, assuming hydrostatic equilibrium, is computed to be $(3 \rightarrow 4) \times 10^{14}\,\text{M}_\odot$ within 0.7 Mpc of the cluster center and $(1 \rightarrow 2) \times 10^{15}\,\text{M}_\odot$ within 3.6 Mpc of the center, consistent with the mass estimate of the virial theorem.

Other clusters of galaxies besides the Coma cluster have had their masses estimated, using the virial theorem applied to their galaxies or the equation of hydrostatic equilibrium applied to their gas. Typical mass-to-light ratios for clusters lie in the range $\langle M/L_B \rangle = 200 \rightarrow 300\,\text{M}_\odot/\text{L}_{\odot,B}$, so the Coma cluster is not unusual in the amount of dark matter which it contains. If the masses of all the clusters of galaxies are added together, it is found that their density parameter is

$$\Omega_{\text{clus},0} \approx 0.2. \tag{8.47}$$

This provides a *lower limit* to the matter density of the universe, since any smoothly distributed matter in the intercluster voids will not be included in this number.

8.4 ■ GRAVITATIONAL LENSING

So far, I have outlined the classical methods for detecting dark matter via its gravitational effects on luminous matter.[7] We can detect dark matter around spiral galaxies because it affects the motions of stars and interstellar gas. We can detect dark matter in clusters of galaxies because it affects the motions of galaxies and intracluster gas. However, as Einstein realized, dark matter will affect not only the trajectory of matter, but also the trajectory of photons. Thus, dark matter can bend and focus light, acting as a *gravitational lens*. The effects of dark matter on photons have been used to search for dark matter within the halo of our own galaxy, as well as in distant clusters of galaxies.

To see how gravitational lensing can be used to detect dark matter, start by considering the dark halo surrounding our galaxy. Some of the dark matter in the halo might consist of massive compact objects such as brown dwarfs, white dwarfs, neutron stars, and black holes. These objects have been collectively called

[7]The roots of these methods can be traced back as far as the year 1846, when Leverrier and Adams deduced the existence of the dim planet Neptune by its effect on the orbit of Uranus.

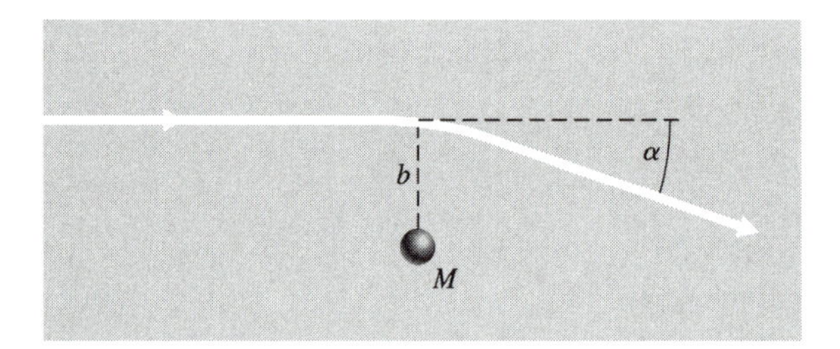

FIGURE 8.5 Deflection of light by a massive compact object.

MACHOs, a slightly strained acronym for MAssive Compact Halo Objects. If a photon passes such a compact massive object at an impact parameter b, as shown in Figure 8.5, the local curvature of space-time will cause the photon to be deflected by an angle

$$\alpha = \frac{4GM}{c^2 b}, \tag{8.48}$$

where M is the mass of the compact object. For instance, light from a distant star that just grazes the Sun's surface should be deflected through an angle

$$\alpha = \frac{4G\,\mathrm{M}_\odot}{c^2\,\mathrm{R}_\odot} = 1.7\,\mathrm{arcsec}. \tag{8.49}$$

In 1919, after Einstein predicted a deflection of this magnitude, an eclipse expedition photographed stars in the vicinity of the Sun. Comparison of the eclipse photographs with photographs of the same star field taken six months earlier revealed that the apparent positions of the stars were deflected by the amount that Einstein had predicted. This result brought fame to Einstein and experimental support to the theory of general relativity.

Since a star, or a brown dwarf, or a stellar remnant, can deflect light, it can act as a lens. Suppose a MACHO in the halo of our galaxy passes directly between an observer in our galaxy and a star in the Large Magellanic Cloud. Figure 8.6 shows such a situation, with a MACHO that happens to be halfway between the observer and the star. As the MACHO deflects the light from the distant star, it produces an image of the star, which is both distorted and amplified. If the MACHO is *exactly* along the line of sight between the observer and the lensed star, the image produced is a perfect ring, with angular radius

$$\theta_E = \left(\frac{4GM}{c^2 d}\frac{1-x}{x}\right)^{1/2}, \tag{8.50}$$

where M is the mass of the lensing MACHO, d is the distance from the observer to the lensed star, and xd (where $0 < x < 1$) is the distance from the observer to the lensing MACHO. The angle θ_E is known as the *Einstein radius*. If $x \approx 0.5$ (that is, if the MACHO is roughly halfway between the observer and the lensed

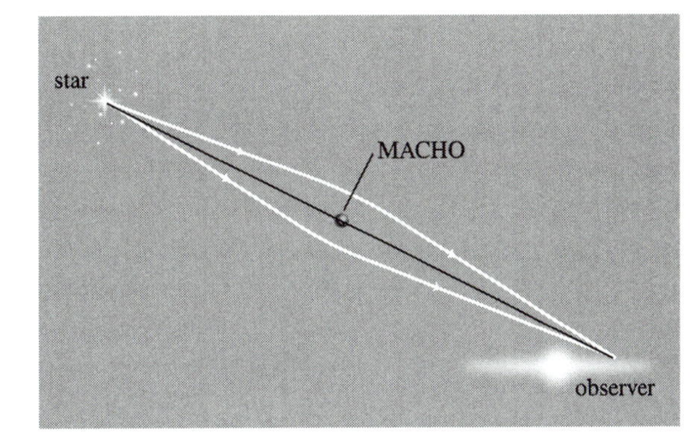

FIGURE 8.6 Light from star in the Large Magellanic Cloud is deflected by a MACHO on its way to an observer in the disk of our galaxy (seen edge-on in this figure).

star), then

$$\theta_E \approx 4 \times 10^{-4}\text{arcsec}\left(\frac{M}{1\,\mathrm{M}_\odot}\right)^{1/2}\left(\frac{d}{50\,\mathrm{kpc}}\right)^{-1/2}. \tag{8.51}$$

If the MACHO does not lie perfectly along the line of sight to the star, then the image of the star will be distorted into two or more arcs instead of a single unbroken ring. Although the Einstein radius for an LMC star being lensed by a MACHO is too small to be resolved, it is possible, in some cases, to detect the amplification of the flux from the star. For the amplification to be significant, the angular distance between the MACHO and the lensed star, as seen from Earth, must be comparable to, or smaller than, the Einstein radius. Given the small size of the Einstein radius, the probability of any particular star in the LMC being lensed at any moment is tiny. It has been calculated that if the dark halo of our galaxy were entirely composed of MACHOs, then the probability of any given star in the LMC being lensed at any given time would still only be $P \sim 5 \times 10^{-7}$.

To detect lensing by MACHOs, various research groups took up the daunting task of monitoring millions of stars in the Large Magellanic Cloud to watch for changes in their flux. Since the MACHOs in our dark halo and the stars in the LMC are in constant relative motion, the typical signature of a "lensing event" is a star that becomes brighter as the angular distance between star and MACHO decreases, then becomes dimmer as the angular distance increases again. The typical time scale for a lensing event is the time it takes a MACHO to travel through an angular distance equal to θ_E as seen from Earth; for a MACHO halfway between here and the LMC, this is

$$\Delta t = \frac{d\,\theta_E}{2v} \approx 90\,\text{days}\left(\frac{M}{1\,\mathrm{M}_\odot}\right)^{1/2}\left(\frac{v}{200\,\mathrm{km\,s^{-1}}}\right)^{-1}, \tag{8.52}$$

where v is the relative transverse velocity of the MACHO and the lensed star as seen by the observer on Earth. Generally speaking, more massive MACHOs produce larger Einstein rings and thus will amplify the lensed star for a longer time.

The research groups that searched for MACHOs found a scarcity of short duration lensing events, suggesting that there is no significant population of brown dwarfs (with $M < 0.08\,\mathrm{M}_\odot$) in the dark halo of our galaxy. The total number of lensing events they detected suggest that as much as 20% of the halo mass could be in the form of MACHOs. The long time scales of the observed lensing events, which have $\Delta t > 35$ days, suggest typical MACHO masses of $M > 0.15\,\mathrm{M}_\odot$. (Perhaps the MACHOs are old, cold white dwarfs, which would have the correct mass.) Alternatively, the observed lensing events could be due, at least in part, to lensing objects within the LMC itself. In any case, the search for MACHOs suggests that most of the matter in the dark halo of our galaxy is due to a smoothly distributed component, instead of being congealed into MACHOs of roughly stellar mass.

Gravitational lensing occurs at all mass scales. Suppose, for instance, that a cluster of galaxies, with $M \sim 10^{14}\,\mathrm{M}_\odot$, at a distance ~ 500 Mpc from our galaxy, lenses a background galaxy at $d \sim 1000$ Mpc. The Einstein radius for this configuration will be

$$\theta_E \approx 0.5\,\mathrm{arcmin}\left(\frac{M}{10^{14}\,\mathrm{M}_\odot}\right)^{1/2}\left(\frac{d}{1000\,\mathrm{Mpc}}\right)^{-1/2}. \quad (8.53)$$

The arc-shaped images into which the background galaxy is distorted by the lensing cluster can thus be resolved. For instance, Figure 8.7 shows an image of the

FIGURE 8.7 A Hubble Space Telescope picture of the rich cluster Abell 2218, displaying gravitationally lensed arcs. The region shown is roughly 2.4 arcmin by 1.2 arcmin, equivalent to 0.54 Mpc by 0.27 Mpc at the distance of Abell 2218 (courtesy of W. Couch [University of New South Wales] and NASA).

cluster Abell 2218, which has a redshift $z = 0.18$, and hence is at a proper distance $d = 770\,\mathrm{Mpc}$. The elongated arcs seen in Figure 8.7 are not oddly shaped galaxies within the cluster; instead, they are background galaxies, at redshifts $z > 0.18$, which are gravitationally lensed by the cluster mass. The mass of clusters can be estimated by the degree to which they lens background galaxies. The masses calculated in this way are in general agreement with the masses found by applying the virial theorem to the motions of galaxies in the cluster or by applying the equation of hydrostatic equilibrium to the hot intracluster gas.

8.5 ■ WHAT'S THE MATTER?

We described how to detect dark matter by its gravitational effects, but have been dodging the essential question: "What is it?" Adding together the masses of clusters of galaxies gives a lower limit on the matter density of the universe, telling us that $\Omega_{m,0} \geq 0.2$. However, the density parameter of baryonic matter is only $\Omega_{\mathrm{bary},0} \approx 0.04$. Thus, the density of nonbaryonic matter is at least four times the density of the familiar baryonic matter of which people and planets and stars are made.

As you might expect, conjecture about the nature of the nonbaryonic dark matter has run rampant (some might even say it has run amok). A component of the universe that is totally invisible is an open invitation to speculation. To give a taste of the variety of speculation, some scientists have proposed that the dark matter might be made of axions, a type of elementary particle with a rest energy of $m_{\mathrm{ax}}c^2 \sim 10^{-5}\,\mathrm{eV}$, equivalent to $m_{\mathrm{ax}} \sim 2 \times 10^{-41}\,\mathrm{kg}$. This is a rather low mass—it would take some 50 billion axions (if they indeed exist) to equal the mass of one electron. On the other hand, some scientists have conjectured that the dark matter might be made of primordial black holes, with masses up to $m_{\mathrm{BH}} \sim 10^5\,\mathrm{M}_\odot$, equivalent to $m_{\mathrm{BH}} \sim 2 \times 10^{35}\,\mathrm{kg}$.[8] This is a rather high mass—it would take some 30 billion Earths to equal the mass of one primordial black hole (if they indeed exist). It is a sign of the vast ignorance concerning nonbaryonic dark matter that these two candidates for the role of dark matter differ in mass by 76 orders of magnitude.

One nonbaryonic particle that we know exists, and which seems to have a nonzero mass, is the neutrino. As stated in section 5.1, there should exist today a cosmic background of neutrinos. Just as the Cosmic Microwave Background is a relic of the time when the universe was opaque to photons, the Cosmic Neutrino Background is a relic of the time when the universe was hot and dense enough to be opaque to neutrinos. The number density of each of the three flavors of neutrinos (ν_e, ν_μ, and ν_τ) has been calculated to be $\frac{3}{11}$ times the number density

[8] A *primordial* black hole is one that formed very early in the history of the universe, rather than by the collapse of a massive star later on.

of CMB photons, yielding a total number density of neutrinos

$$n_\nu = 3\left(\frac{3}{11}\right)n_\gamma = \left(\frac{9}{11}\right)(4.11 \times 10^8\,\mathrm{m}^{-3}) = 3.36 \times 10^8\,\mathrm{m}^{-3}. \tag{8.54}$$

This means that at any moment, about twenty million cosmic neutrinos are zipping through your body, "like photons through a pane of glass." In order to provide *all* the nonbaryonic mass in the universe, the average neutrino mass would have to be

$$m_\nu c^2 = \frac{\Omega_{\mathrm{dm},0}\varepsilon_{c,0}}{n_\nu}. \tag{8.55}$$

Given a density parameter in nonbaryonic dark matter of $\Omega_{\mathrm{dm},0} \approx 0.26$, this implies that a mean neutrino mass of

$$m_\nu c^2 \approx \frac{0.26(5200\,\mathrm{MeV\,m^{-3}})}{3.36 \times 10^8\,\mathrm{m}^{-3}} \approx 4\,\mathrm{eV} \tag{8.56}$$

would be necessary to provide all the nonbaryonic dark matter in the universe.

Evidence indicates that neutrinos do have some mass. But how much? Enough to contribute significantly to the energy density of the universe? The observations of neutrinos from the Sun, as mentioned in section 2.4, indicate that electron neutrinos oscillate into some other flavor of neutrino, with the difference in the squares of the masses of the two neutrinos being $\Delta(m_\nu^2 c^4) \approx 5 \times 10^{-5}\,\mathrm{eV}^2$. Observations of muon neutrinos created in the Earth's atmosphere indicate that muon neutrinos oscillate into tau neutrinos, with $\Delta(m_\nu^2 c^4) \approx 3 \times 10^{-3}\,\mathrm{eV}^2$. The minimum neutrino masses consistent with these results would have one flavor with $m_\nu c^2 \sim 0.05\,\mathrm{eV}$, another with $m_\nu c^2 \sim 0.007\,\mathrm{eV}$, and the third with $m_\nu c^2 \ll 0.007\,\mathrm{eV}$. If the neutrino masses are this small, then the density parameter in neutrinos is only $\Omega_\nu \sim 10^{-3}$, and neutrinos make up less than 0.5% of the nonbaryonic dark matter. If, on the other hand, neutrinos make up all the nonbaryonic dark matter, the masses of the three species would have to be very nearly identical; for instance, one neutrino flavor with $m_\nu c^2 = 4.0\,\mathrm{eV}$, another with $m_\nu c^2 = 4.0004\,\mathrm{eV}$, and the third with $m_\nu c^2 = 4.000006\,\mathrm{eV}$ would be in agreement with the deduced values of $\Delta(m_\nu^2 c^4)$.

If the masses of all three neutrinos turn out to be significantly less than $m_\nu c^2 \sim 4\,\mathrm{eV}$, then the bulk of the nonbaryonic dark matter in the universe must be made of some particle other than neutrinos. Particle physicists have provided several possible candidates for the role of dark matter. For instance, consider the extension of the Standard Model of particle physics known as supersymmetry. Various supersymmetric models predict the existence of massive nonbaryonic particles such as photinos, gravitinos, axinos, sneutrinos, gluinos, and so forth. The fact that none of these "inos" have been seen in particle accelerator experiments means that they must be massive (if they exist), with $mc^2 > 10\,\mathrm{GeV}$.

Like neutrinos, the hypothetical supersymmetric particles interact with other particles only through gravity and through the weak nuclear force, which makes them intrinsically difficult to detect. Particles that interact via the weak nuclear force, but which are much more massive than the upper limit on the neutrino mass, are known generically as Weakly Interacting Massive Particles, or WIMPs.[9] Since WIMPs, like neutrinos, do interact with atomic nuclei on occasion, experimenters have set up WIMP detectors to discover cosmic WIMPs. So far, no convincing detections have been made—but the search goes on.

SUGGESTED READING

Full references are given in the Annotated Bibliography on page 235.

Liddle (1999), ch. 8: A brief sketch of methods for detecting dark matter.

Peacock (1999), ch. 12: Dark matter in the universe, both baryonic and nonbaryonic. Also, chapter 4 gives a good review of gravitational lensing.

Rich (2001), ch. 2.4: A discussion of the dark matter candidates.

PROBLEMS

8.26. Suppose it were suggested that black holes of mass $10^{-8}\,M_\odot$ made up all the dark matter in the halo of our galaxy. How far away would you expect the nearest such black hole to be? How frequently would you expect such a black hole to pass within 1 AU of the Sun? (An order-of-magnitude estimate is sufficient.)

Suppose it were suggested that MACHOs of mass $10^{-3}\,M_\odot$ (about the mass of Jupiter) made up all the dark matter in the halo of our galaxy. How far away would you expect the nearest MACHO to be? How frequently would such a MACHO pass within 1 AU of the Sun? (Again, an order-of-magnitude estimate will suffice.)

8.27. The Draco galaxy is a dwarf galaxy within the Local Group. Its luminosity is $L = (1.8 \pm 0.8) \times 10^5\,L_\odot$ and half its total luminosity is contained within a sphere of radius $r_h = 120 \pm 12\,\text{pc}$. The red giant stars in the Draco galaxy are bright enough to have their line-of-sight velocities measured. The measured velocity dispersion of the red giant stars in the Draco galaxy is $\sigma_r = 10.5 \pm 2.2\,\text{km}\,\text{s}^{-1}$. What is the mass of the Draco galaxy? What is its mass-to-light ratio? Describe the possible sources of error in your mass estimate of this galaxy.

8.28. A light ray just grazes the surface of the Earth ($M = 6.0 \times 10^{24}\,\text{kg}$, $R = 6.4 \times 10^6\,\text{m}$). Through what angle α is the light ray bent by gravitational lensing? (Ignore the refractive effects of the Earth's atmosphere.) Repeat your calculation for a white dwarf ($M = 2.0 \times 10^{30}\,\text{kg}$, $R = 1.5 \times 10^7\,\text{m}$) and for a neutron star ($M = 3.0 \times 10^{30}\,\text{kg}$, $R = 1.2 \times 10^4\,\text{m}$).

8.29. If the halo of our galaxy is spherically symmetric, what is the mass density $\rho(r)$ within the halo? If the universe contains a cosmological constant with density parameter

[9] The acronym "MACHO," encountered in the previous section, was first coined as a humorous riposte to the acronym "WIMP."

$\Omega_{\Lambda,0} = 0.7$, would you expect it to significantly affect the dynamics of our galaxy's halo? Explain why or why not.

8.30. In the previous chapter, we noted that galaxies in rich clusters are poor standard candles, because they tend to grow brighter with time as they merge with other galaxies. Let's estimate the galaxy merger rate in the Coma cluster to see whether it's truly significant. The Coma cluster contains $N \approx 1000$ galaxies within its half-mass radius of $r_h \approx 1.5\,\text{Mpc}$. What is the mean number density of galaxies within the half-mass radius? Suppose that the typical cross-section of a galaxy is $\Sigma \approx 10^{-3}\,\text{Mpc}^2$. How far will a galaxy in the Coma cluster travel, on average, before it collides with another galaxy? The velocity dispersion of the Coma cluster is $\sigma \approx 880\,\text{km}\,\text{s}^{-3}$. What is the average time between collisions for a galaxy in the Coma cluster? Is this time greater than or less than the Hubble time?

CHAPTER 9

The Cosmic Microwave Background

If Heinrich Olbers had lived in intergalactic space and had eyes that operated at millimeter wavelengths (admittedly a very large "if"), he would not have formulated Olbers' Paradox. At wavelengths of a few millimeters, thousands of times longer than human eyes can detect, most of the light in the universe comes not from the hot balls of gas we call stars, but from the Cosmic Microwave Background (CMB). Unknown to Olbers, the night sky actually *is* uniformly bright—it's just uniformly bright at a temperature of $T_0 = 2.725\,\text{K}$ rather than at a temperature of a few thousand degrees Kelvin. The current energy density of the Cosmic Microwave Background,

$$\varepsilon_{\gamma,0} = \alpha T_0^4 = 0.261\,\text{MeV}\,\text{m}^{-3}, \tag{9.1}$$

is only 5×10^{-5} times the current critical density. However, since the energy per CMB photon is small ($hf_{\text{mean}} = 6.34 \times 10^{-4}\,\text{eV}$), the number density of CMB photons in the universe is large:

$$n_{\gamma,0} = 4.11 \times 10^8\,\text{m}^{-3}. \tag{9.2}$$

It is particularly enlightening to compare the energy density and number density of photons to those of baryons (that is, protons and neutrons). Given a current density parameter for baryons of $\Omega_{\text{bary},0} \approx 0.04$, the current energy density of baryons is

$$\varepsilon_{\text{bary},0} = \Omega_{\text{bary},0}\varepsilon_{c,0} \approx 0.04(5200\,\text{MeV}\,\text{m}^{-3}) \approx 210\,\text{MeV}\,\text{m}^{-3}. \tag{9.3}$$

Thus, the energy density in baryons today is about 800 times the energy density in CMB photons. Note, though, that the rest energy of a proton or neutron is $E_{\text{bary}} \approx 939\,\text{MeV}$; this is more than a trillion times the mean energy of a CMB photon. The number density of baryons, therefore, is much lower than the number density of photons:

$$n_{\text{bary},0} = \frac{\varepsilon_{\text{bary},0}}{E_{\text{bary}}} \approx \frac{210\,\text{MeV}\,\text{m}^{-3}}{939\,\text{MeV}} \approx 0.22\,\text{m}^{-3}. \tag{9.4}$$

The ratio of baryons to photons in the universe (a number usually designated by the Greek letter η) is, from equations (9.2) and (9.4),

$$\eta = \frac{n_{\text{bary},0}}{n_{\gamma,0}} \approx \frac{0.22\,\text{m}^{-3}}{4.11 \times 10^8\,\text{m}^{-3}} \approx 5 \times 10^{-10}. \tag{9.5}$$

Baryons are badly outnumbered by photons in the universe as a whole, by a ratio of roughly two billion to one.

9.1 ■ OBSERVING THE CMB

Although CMB photons are as common as dirt,[1] Arno Penzias and Robert Wilson were surprised when they serendipitously discovered the Cosmic Microwave Background. At the time of their discovery, Penzias and Wilson were radio astronomers working at Bell Laboratories. The horn-reflector radio antenna they used had previously been utilized to receive microwave signals, of wavelength $\lambda = 7.35\,\text{cm}$, reflected from an orbiting communications satellite. Turning from telecommunications to astronomy, Penzias and Wilson found a slightly stronger signal than they expected when they turned the antenna toward the sky. They did everything they could think of to reduce "noise" in their system. They even shooed away a pair of pigeons that had roosted in the antenna and cleaned up what they later called "the usual white dielectric" generated by pigeons.

The excess signal remained. It was isotropic and constant with time, so it couldn't be associated with an isolated celestial source. Wilson and Penzias were puzzled until they were put in touch with Robert Dicke and his research group at Princeton University. Dicke had deduced that the universe, if it started in a hot dense state, should now be filled with microwave radiation.[2] In fact, Dicke and his group were in the process of building a microwave antenna when Penzias and Wilson told them that they had already detected the predicted microwave radiation. Penzias and Wilson wrote a paper for *The Astrophysical Journal* in which they wrote, "Measurements of the effective zenith noise temperature of the 20-foot horn-reflector antenna . . . at 4080 Mc/s have yielded a value about 3.5 K higher than expected. This excess temperature is, within the limits of our observations, isotropic, unpolarized, and free from seasonal variations (July, 1964–April, 1965). A possible explanation for the observed excess noise temperature is the one given by Dicke, Peebles, Roll, and Wilkinson in a companion letter in this issue." The companion paper by Dicke and his collaborators points out that the radiation could be a relic of an early, hot, dense, and opaque state of the universe.

Measuring the spectrum of the CMB and confirming that it is indeed a blackbody, is not a simple task, even with modern technology. The current energy per

[1] Actually, much commoner than dirt, when you stop to think of it, since dirt is made of baryons.

[2] To give credit where it's due, the existence of the cosmic background radiation had actually been predicted by George Gamow, Ralph Alpher, and Robert Herman as early as 1948; unfortunately, the prediction wasn't acted on at the time, and had fallen into obscurity during the intervening years.

CMB photon, $\sim 6 \times 10^{-4}$ eV, is tiny compared to the energy required to break up an atomic nucleus ($\sim$ 1 MeV) or even the energy required to ionize an atom ($\sim$ 10 eV). However, the mean photon energy is comparable to the energy of vibration or rotation for a small molecule such as H_2O. Thus, CMB photons can zip along for more than 13 billion years through the tenuous intergalactic medium, then be absorbed a microsecond away from the Earth's surface by a water molecule in the atmosphere. Microwaves with wavelengths shorter than $\lambda \sim 3$ cm are strongly absorbed by water molecules. Penzias and Wilson observed the CMB at a wavelength $\lambda = 7.35$ cm because that was the wavelength of the signals that Bell Labs had been bouncing off orbiting satellites. Thus, Penzias and Wilson were observing at a wavelength 40 times longer than the wavelength ($\lambda \approx 2$ mm) at which the CMB spectrum reaches its peak.

The CMB can be measured at wavelengths shorter than 3 cm by observing from high-altitude balloons or from the South Pole, where the combination of cold temperatures and high altitude[3] keeps the atmospheric humidity low. The best way to measure the spectrum of the CMB, however, is to go completely above the damp atmosphere of the Earth. The CMB spectrum was first measured accurately over a wide range of wavelengths by the COsmic Background Explorer (COBE) satellite, launched in 1989, into an orbit 900 km above the Earth's surface. COBE actually contained three different instruments. The Diffuse InfraRed Background Experiment (DIRBE) was designed to measure radiation at the wavelengths $0.001\text{ mm} < \lambda < 0.24\text{ mm}$; at these wavelengths, it was primarily detecting stars and dust within our own galaxy. The second instrument, called the Far InfraRed Absolute Spectrophotometer (FIRAS), was used to measure the spectrum of the CMB in the range $0.1\text{ mm} < \lambda < 10\text{ mm}$, a wavelength band that includes the peak in the CMB spectrum. The third instrument, called the Differential Microwave Radiometer (DMR), was designed to make full-sky maps of the CMB at three different wavelengths: $\lambda = 3.3$ mm, 5.7 mm, and 9.6 mm. Three important results came from the analysis of the COBE data.

Result number one: At any angular position (θ, ϕ) on the sky, the spectrum of the Cosmic Microwave Background is very close to that of an ideal blackbody, as illustrated in Figure 9.1. How close is very close? FIRAS could have detected fluctuations in the spectrum as small as $\Delta\epsilon/\epsilon \approx 10^{-4}$. No deviations were found at this level within the wavelength range investigated by FIRAS.

Result number two: The CMB has the *dipole* distortion in temperature shown in the top panel of Figure 9.2.[4] That is, although each point on the sky has a blackbody spectrum, in one half of the sky the spectrum is slightly blueshifted to higher temperatures, and in the other half the spectrum is slightly redshifted to lower temperatures.[5] This dipole distortion is a simple Doppler shift, caused

[3] The South Pole is nearly 3 kilometers above sea level.

[4] The dipole distortion of the CMB was first detected in 1977, using aircraft-borne and balloon-borne detectors. The unique contribution of COBE was the precision with which it measured the temperature distortion.

[5] The distorted "yin-yang" pattern in the upper panel of Figure 9.2 represents the darker, cooler (yin?) hemisphere of the sky and the hotter, brighter (yang?) hemisphere, distorted by the map projection.

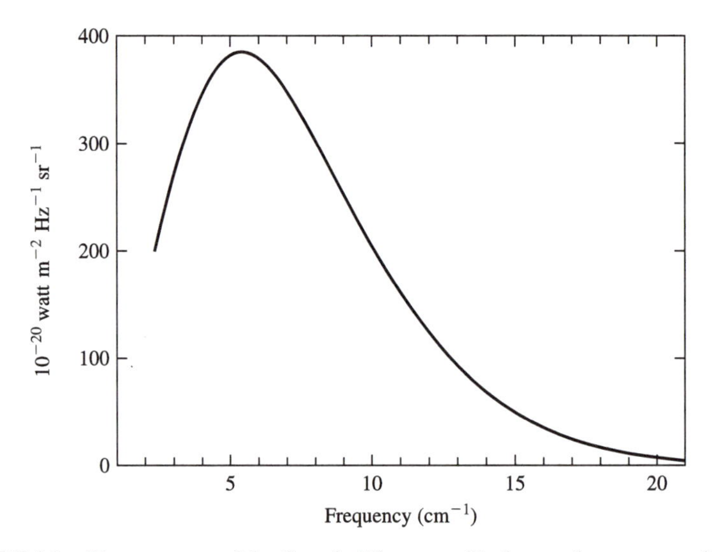

FIGURE 9.1 The spectrum of the Cosmic Microwave Background, as measured by the FIRAS instrument on the COBE satellite. The uncertainties in the measurement are smaller than the thickness of the line.

by the net motion of the COBE satellite relative to a frame of reference in which the CMB is isotropic. After correcting for the orbital motion of COBE around the Earth ($v \sim 8\,\text{km}\,\text{s}^{-1}$), for the orbital motion of the Earth around the Sun ($v \sim 30\,\text{km}\,\text{s}^{-1}$), for the orbital motion of the Sun around the galactic center ($v \sim 220\,\text{km}\,\text{s}^{-1}$), and for the orbital motion of our galaxy relative to the center of mass of the Local Group ($v \sim 80\,\text{km}\,\text{s}^{-1}$), it is found that the Local Group is moving in the general direction of the constellation Hydra, with a speed $v_{\text{LG}} = 630 \pm 20\,\text{km}\,\text{s}^{-1} = 0.0021c$. This peculiar velocity for the Local Group is what you'd expect as the result of gravitational acceleration by the largest lumps of matter in the vicinity of the Local Group. The Local Group is being accelerated toward the Virgo cluster, the nearest big cluster to us. In addition, the Virgo cluster is being accelerated toward the Hydra-Centaurus supercluster, the nearest supercluster to us. The combination of these two accelerations, working over the age of the universe, has launched the Local Group in the direction of Hydra, at 0.2% of the speed of light.

Result number three: After the dipole distortion of the CMB is subtracted away, the remaining temperature fluctuations, shown in the lower panel of Figure 9.2, are small in amplitude. Let the temperature of the CMB, at a given point on the sky, be $T(\theta, \phi)$. The mean temperature, averaging over all locations, is

$$\langle T \rangle = \frac{1}{4\pi} \int T(\theta, \phi) \sin\theta \, d\theta \, d\phi = 2.725\,\text{K}. \tag{9.6}$$

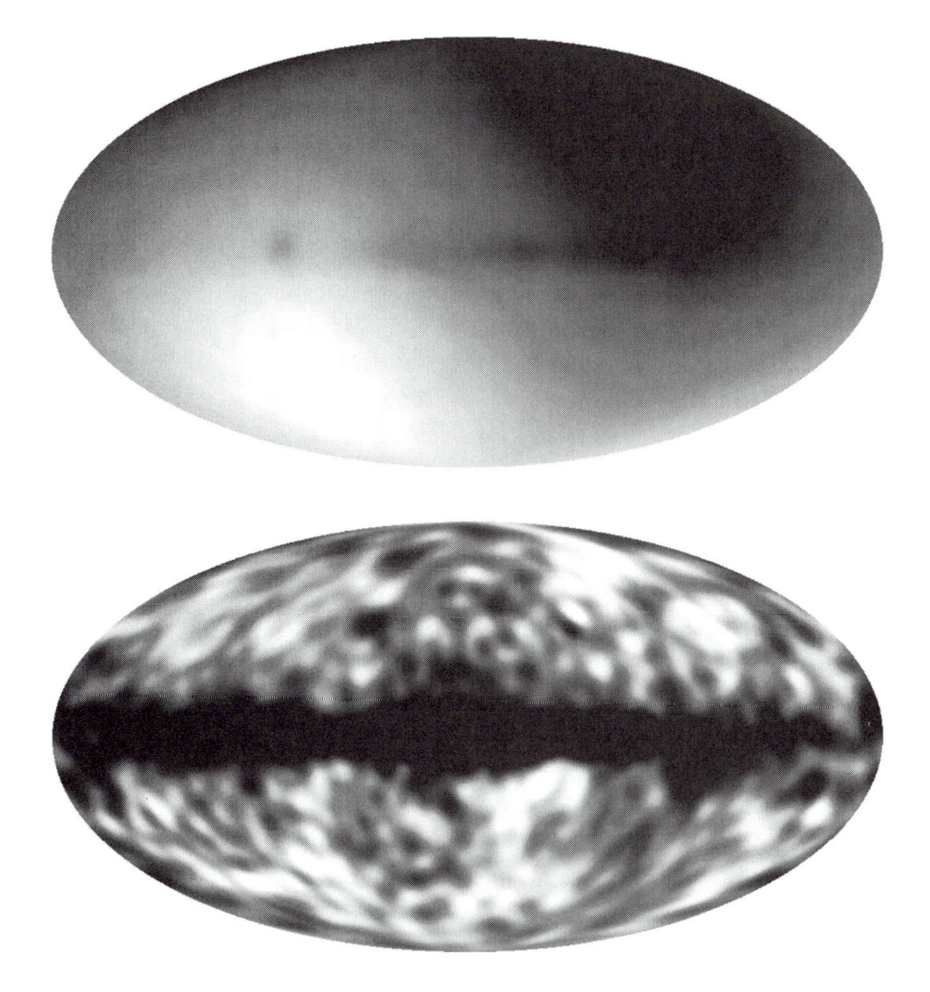

FIGURE 9.2 Top panel: The fluctuations in temperature in the CMB, as measured by COBE. Bottom panel: The fluctuations in temperature remaining after subtraction of the dipole due to the satellite's proper motion. The band across the middle is due to emission from the disk of our own galaxy. (Courtesy of NASA Goddard Space Flight Center and the COBE Science Working Group.)

The dimensionless temperature fluctuation at a given point on the sky is

$$\frac{\delta T}{T}(\theta, \phi) \equiv \frac{T(\theta, \phi) - \langle T \rangle}{\langle T \rangle}. \tag{9.7}$$

From the maps of the sky made by the DMR instrument aboard COBE, it was found that after subtraction of the Doppler dipole, the root mean square temperature fluctuation was

$$\left\langle \left(\frac{\delta T}{T} \right)^2 \right\rangle^{1/2} = 1.1 \times 10^{-5}. \tag{9.8}$$

(This analysis excludes the regions of the sky contaminated by foreground emission from our own galaxy.) The fact that the temperature of the CMB varies by only 30 microKelvin across the sky represents a remarkably close approach to isotropy.[6]

The observations that the CMB has a nearly perfect blackbody spectrum and that it is nearly isotropic (once the Doppler dipole is removed) provide strong support for the Hot Big Bang model of the universe. A background of nearly isotropic blackbody radiation is natural if the universe was once hot, dense, opaque, and nearly homogeneous, as it was in the Hot Big Bang scenario. If the universe did not go through such a phase, then any explanation of the Cosmic Microwave Background will have to be much more contrived.

9.2 ■ RECOMBINATION AND DECOUPLING

To understand in more detail the origin of the Cosmic Microwave Background, we'll have to examine fairly carefully the process by which the baryonic matter goes from being an ionized plasma to a gas of neutral atoms, and the closely related process by which the universe goes from being opaque to being transparent. To avoid muddle, we will distinguish among three closely related (but not identical) moments in the history of the universe. First, the epoch of *recombination* is the time at which the baryonic component of the universe goes from being ionized to being neutral. Numerically, we might define it as the instant in time when the number density of ions is equal to the number density of neutral atoms.[7] Second, the epoch of *photon decoupling* is the time when the rate at which photons scatter from electrons becomes smaller than the Hubble parameter (which tells us the rate at which the universe expands). When photons decouple, they cease to interact with the electrons, and the universe becomes transparent. Third, the epoch of *last scattering* is the time at which a typical CMB photon underwent its last scattering from an electron. Surrounding every observer in the universe is a *last scattering surface*, illustrated in Figure 9.3, from which the CMB photons have been streaming freely, with no further scattering by electrons. The probability that a photon will scatter from an electron is small once the expansion rate of the universe is faster than the scattering rate; thus, the epoch of last scattering is very close to the epoch of photon decoupling.

To keep things from getting too complicated, we will assume that the baryonic component of the universe consisted entirely of hydrogen at the epoch of recom-

[6]To make an analogy, if the surface of the Earth were smooth to 11 parts per million, the highest mountains would be just seventy meters above the deepest ocean trenches.

[7]Cosmologists sometimes grumble that this should really be called the epoch of "combination" rather than the epoch of "recombination," since this is the very first time when electrons and ions combined to form neutral atoms.

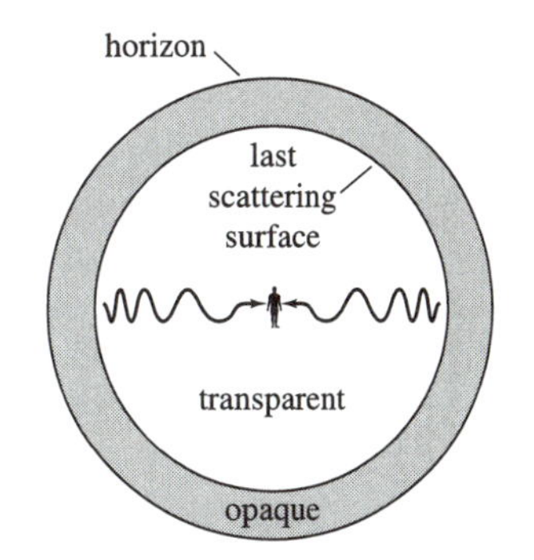

FIGURE 9.3 An observer is surrounded by a spherical last scattering surface. The photons of the CMB travel straight to us from the last scattering surface, being continuously redshifted.

bination. This is not, however, a strictly accurate assumption. Even at the time of recombination, before stars had a chance to pollute the universe with heavy elements, there was a significant amount of helium present.[8] However, the presence of helium is merely a complicating factor. All the significant physics of recombination can be studied in a simplified universe containing no elements other than hydrogen. The hydrogen can take the form of a neutral atom (designated by the letter H), or of a naked hydrogen nucleus, otherwise known as a proton (designated by the letter p). To maintain charge neutrality in this hydrogen-only universe, the number density of free electrons must be equal to that of free protons: $n_e = n_p$. The degree to which the baryonic content of the universe is ionized can be expressed as the fractional ionization X, defined as

$$X \equiv \frac{n_p}{n_p + n_{\rm H}} = \frac{n_p}{n_{\rm bary}} = \frac{n_e}{n_{\rm bary}}. \tag{9.9}$$

The value of X ranges from $X = 1$, when the baryonic content is fully ionized, to $X = 0$, when it consists entirely of neutral atoms.

One useful consequence of assuming that hydrogen is the only element is that there is now a single relevant energy scale in the problem: the ionization energy of hydrogen, $Q = 13.6\,\text{eV}$. A photon with an energy $hf > Q$ is capable of photoionizing a hydrogen atom:

$$\text{H} + \gamma \rightarrow p + e^-. \tag{9.10}$$

This reaction can run in the opposite direction, as well; a proton and an electron can undergo *radiative recombination*, forming a bound hydrogen atom while a photon carries away the excess energy:

$$p + e^- \rightarrow \text{H} + \gamma. \tag{9.11}$$

[8]In the next chapter, we will examine how and why this helium was formed in the early universe.

In a universe containing protons, electrons, and photons, the fractional ionization X will depend on the balance between photoionization and radiative recombination.

Let's travel back in time to a period before the epoch of recombination. For concreteness, let's choose the moment when $a = 10^{-5}$, corresponding to a redshift $z = 10^5$. (In the Benchmark Model, this scale factor was reached when the universe was seventy years old.) The temperature of the background radiation at this time was $T \approx 3 \times 10^5$ K, and the average photon energy was $hf_{\rm mean} \approx 2.7kT \approx 60$ eV. With such a high energy per photon, and with a ratio of photons to baryons of nearly two billion, any hydrogen atoms that happened to form by radiative recombination were very short-lived; almost immediately, they were blasted apart into their component electron and proton by a high-energy photon. At early times, then, the fractional ionization of the universe was very close to $X = 1$.

When the universe was fully ionized, photons interacted primarily with electrons, and the main interaction mechanism was Thomson scattering:

$$\gamma + e^- \rightarrow \gamma + e^- . \tag{9.12}$$

The scattering interaction is accompanied by a transfer of energy and momentum between the photon and electron. The cross-section for Thomson scattering is $\sigma_e = 6.65 \times 10^{-29}\,\mathrm{m}^2$. The mean free path of a photon—that is, the mean distance it travels before scattering from an electron—is

$$\lambda = \frac{1}{n_e \sigma_e} . \tag{9.13}$$

Since photons travel with a speed c, the rate at which a photon undergoes scattering interactions is

$$\Gamma = \frac{c}{\lambda} = n_e \sigma_e c . \tag{9.14}$$

When the baryonic component of the universe is fully ionized, $n_e = n_p = n_{\rm bary}$. Currently, the number density of baryons is $n_{\rm bary,0} = 0.22\,\mathrm{m}^{-3}$. The number density of conserved particles, such as baryons, goes as $1/a^3$, so when the early universe was fully ionized, the free electron density was

$$n_e = n_{\rm bary} = \frac{n_{\rm bary,0}}{a^3} , \tag{9.15}$$

and the scattering rate for photons was

$$\Gamma = \frac{n_{\rm bary,0}\sigma_e c}{a^3} = \frac{4.4 \times 10^{-21}\,\mathrm{s}^{-1}}{a^3} . \tag{9.16}$$

This means, for instance, that at $a = 10^{-5}$, photons would scatter from electrons at a rate $\Gamma = 4.4 \times 10^{-6}\,\mathrm{s}^{-1}$, about three times a week.

The photons remain coupled to the electrons as long as their scattering rate, Γ, is larger than H, the rate at which the universe expands; this is equivalent to saying that their mean free path λ is shorter than the Hubble distance c/H. As long as photons scatter frequently from electrons, the photons remain in thermal equilibrium with the electrons (and, indirectly, with the protons as well, thanks to the electrons' interactions with the protons). The photons, electrons, and protons, as long as they remain in thermal equilibrium, all have the same temperature T. When the photon scattering rate Γ drops below H, then the electrons are being diluted by expansion more rapidly than the photons can interact with them. The photons then decouple from the electrons and the universe becomes transparent. Once the photons are decoupled from the electrons and protons, the baryonic portion of the universe is no longer compelled to have the same temperature as the Cosmic Microwave Background. During the early stages of the universe ($a < a_{rm} \approx 3 \times 10^{-4}$) the universe was radiation dominated, and the Friedmann equation was

$$\frac{H^2}{H_0^2} = \frac{\Omega_{r,0}}{a^4}. \tag{9.17}$$

Thus, the Hubble parameter was

$$H = \frac{H_0 \Omega_{r,0}^{1/2}}{a^2} = \frac{2.1 \times 10^{-20}\,\mathrm{s}^{-1}}{a^2}. \tag{9.18}$$

This means, for instance, that at $a = 10^{-5}$, the Hubble parameter was $H = 2.1 \times 10^{-10}\,\mathrm{s}^{-1}$. Since this is much smaller than the scattering rate $\Gamma = 4.4 \times 10^{-6}\,\mathrm{s}^{-1}$ at the same scale factor, the photons were well coupled to the electrons and protons.

If hydrogen remained ionized (and note the qualifying *if*), then photons would have remained coupled to the electrons and protons until a relatively recent time. Taking into account the transition from a radiation-dominated to a matter-dominated universe, and the resulting change in the expansion rate, we can compute that *if* hydrogen had remained fully ionized, then decoupling would have taken place at a scale factor $a \approx 0.023$, corresponding to a redshift $z \approx 42$ and a CMB temperature of $T \approx 120\,\mathrm{K}$. However, at such a low temperature, the CMB photons are too low in energy to keep the hydrogen ionized. Thus, the decoupling of photons is not a gradual process, caused by the continuous lowering of free electron density as the universe expands. Rather, it is a relatively sudden process, caused by the abrupt plummeting of free electron density during the epoch of recombination, as electrons combined with protons to form hydrogen atoms.

9.3 ■ THE PHYSICS OF RECOMBINATION

When does recombination, and the consequent photon decoupling, take place? It's easy to do a quick and dirty approximation of the recombination temperature. Recombination, one could argue, must take place when the mean energy per photon of the Cosmic Microwave Background falls below the ionization energy of hydrogen, $Q = 13.6\,\mathrm{eV}$. When this happens, the average CMB photon is no longer able to photoionize hydrogen. Since the mean CMB photon energy is $\sim 2.7kT$, this line of argument would indicate a recombination temperature of

$$T_{\rm rec} \sim \frac{Q}{2.7k} \sim \frac{13.6\,\mathrm{eV}}{2.7(8.6 \times 10^{-5}\,\mathrm{eV\,K^{-1}})} \sim 60{,}000\,\mathrm{K}. \tag{9.19}$$

Alas, this crude approximation is a little *too* crude to be useful. It doesn't take into account the fact that CMB photons are not of uniform energy—a blackbody spectrum has an exponential tail (see Figure 2.7) trailing off to high energies. Although the mean photon energy is $2.7kT$, about one photon in 500 will have $E > 10kT$, one in 3 million will have $E > 20kT$, and one in 30 billion will have $E > 30kT$. Although extremely high energy photons make up only a tiny fraction of the CMB photons, the total number of CMB photons is enormous—nearly 2 billion photons for every baryon. The vast swarms of photons that surround every newly formed hydrogen atom greatly increase the probability that the atom will collide with a photon from the high-energy tail of the blackbody spectrum, and be photoionized.

Thus, we expect the recombination temperature to depend on the baryon-to-photon ratio η as well as on the ionization energy Q. An exact calculation of the fractional ionization X, as a function of η and T, requires a smattering of statistical mechanics. Let's start with the reaction that determines the value of X in the early universe:

$$\mathrm{H} + \gamma \rightleftharpoons p + e^-. \tag{9.20}$$

While the photons are still coupled to the baryonic component, this reaction will be in statistical equilibrium, with the photoionization rate (going from left to right) balancing the radiative recombination rate (going from right to left). When a reaction is in statistical equilibrium at a temperature T, the number density n_x of particles with mass m_x is given by the Maxwell–Boltzmann equation

$$n_x = g_x \left(\frac{m_x kT}{2\pi\hbar^2}\right)^{3/2} \exp\left(-\frac{m_x c^2}{kT}\right), \tag{9.21}$$

as long as the particles are nonrelativistic, with $kT \ll m_x c^2$. In equation (9.21), g_x is the statistical weight of particle x. For instance, electrons, protons, and neutrons (and their anti-particles as well) all have a statistical weight $g_x = 2$, corresponding to their two possible spin states.[9] From the Maxwell–Boltzmann equation for

[9] If you are a true thermodynamic maven, you will have noted that equation (9.21) omits the chemical potential term, μ, which appears in the most general form of the Maxwell–Boltzmann equation. In most cosmological contexts, as it turns out, the chemical potential is small enough to be safely neglected.

H, p, and e^-, we can construct an equation that relates the number densities of these particles:

$$\frac{n_{\rm H}}{n_p n_e} = \frac{g_{\rm H}}{g_p g_e}\left(\frac{m_{\rm H}}{m_p m_e}\right)^{3/2}\left(\frac{kT}{2\pi\hbar^2}\right)^{-3/2}\exp\left(\frac{[m_p + m_e - m_{\rm H}]c^2}{kT}\right). \quad (9.22)$$

Equation (9.22) can be simplified further. First, since the mass of an electron is small compared to that of a proton, we can set $m_{\rm H}/m_p = 1$. Second, the binding energy $Q = 13.6\,{\rm eV}$ is given by the formula $(m_p + m_e - m_{\rm H})c^2 = Q$. The statistical weights of the proton and electron are $g_p = g_e = 2$, while the statistical weight of a hydrogen atom is $g_{\rm H} = 4$. Thus, the factor $g_{\rm H}/(g_p g_e)$ can be set equal to one. The resulting equation,

$$\frac{n_{\rm H}}{n_p n_e} = \left(\frac{m_e kT}{2\pi\hbar^2}\right)^{-3/2}\exp\left(\frac{Q}{kT}\right), \quad (9.23)$$

is called the *Saha equation*. Our next job is to convert the Saha equation into a relation among X, T, and η. From the definition of X (equation (9.9)), we can make the substitution

$$n_{\rm H} = \frac{1-X}{X} n_p, \quad (9.24)$$

and from the requirement of charge neutrality, we can make the substitution $n_e = n_p$. This yields

$$\frac{1-X}{X} = n_p\left(\frac{m_e kT}{2\pi\hbar^2}\right)^{-3/2}\exp\left(\frac{Q}{kT}\right). \quad (9.25)$$

To eliminate n_p from the above equation, we recall that $\eta \equiv n_{\rm bary}/n_\gamma$. In a universe where hydrogen is the only element, and a fraction X of the hydrogen is in the form of naked protons, we may write

$$\eta = \frac{n_p}{X n_\gamma}. \quad (9.26)$$

Since the photons have a blackbody spectrum, for which

$$n_\gamma = \frac{2.404}{\pi^2}\left(\frac{kT}{\hbar c}\right)^3 = 0.243\left(\frac{kT}{\hbar c}\right)^3, \quad (9.27)$$

we can combine equations (9.26) and (9.27) to find

$$n_p = 0.243 X\eta\left(\frac{kT}{\hbar c}\right)^3. \quad (9.28)$$

Substituting equation (9.28) back into equation (9.25), we finally find the desired equation for X in terms of T and η:

$$\frac{1-X}{X^2} = 3.84\eta\left(\frac{kT}{m_e c^2}\right)^{3/2}\exp\left(\frac{Q}{kT}\right). \tag{9.29}$$

This is a quadratic equation in X, whose positive root is

$$X = \frac{-1+\sqrt{1+4S}}{2S}, \tag{9.30}$$

where

$$S(T,\eta) = 3.84\eta\left(\frac{kT}{m_e c^2}\right)^{3/2}\exp\left(\frac{Q}{kT}\right). \tag{9.31}$$

If we define the moment of recombination as the exact instant when $X = \frac{1}{2}$, then (assuming $\eta = 5.5 \times 10^{-10}$) the recombination temperature is

$$kT_{\rm rec} = 0.323\,{\rm eV} = \frac{Q}{42}. \tag{9.32}$$

Because of the exponential dependence of S upon the temperature, the exact value of η doesn't strongly affect the value of $T_{\rm rec}$. In degrees Kelvin, $kT_{\rm rec} = 0.323\,{\rm eV}$ corresponds to a temperature $T_{\rm rec} = 3740\,{\rm K}$, slightly higher than the melting point of tungsten.[10] The temperature of the universe had a value $T = T_{\rm rec} = 3740\,{\rm K}$ at a redshift $z_{\rm rec} = 1370$, when the age of the universe, in the Benchmark Model, was $t_{\rm rec} = 240{,}000\,{\rm yr}$. Recombination was not an instantaneous process; however, as shown in Figure 9.4, it proceeded fairly rapidly. The fractional ionization goes from $X = 0.9$ at a redshift $z = 1475$ to $X = 0.1$ at a redshift $z = 1255$. In the Benchmark Model, the time that elapses from $X = 0.9$ to $X = 0.1$ is $\Delta t \approx 70{,}000\,{\rm yr}$.

Since the number density of free electrons drops rapidly during the epoch of recombination, the time of photon decoupling comes soon after the time of recombination. The rate of photon scattering, when the hydrogen is partially ionized, is

$$\Gamma(z) = n_e(z)\sigma_e c = X(z)(1+z)^3 n_{\rm bary,0}\sigma_e c. \tag{9.33}$$

Using $\Omega_{\rm bary,0} = 0.04$, the numerical value of the scattering rate is

$$\Gamma(z) = 4.4 \times 10^{-21}\,{\rm s}^{-1} X(z)(1+z)^3. \tag{9.34}$$

While recombination is taking place, the universe is matter-dominated, so the Hubble parameter is given by the relation

$$\frac{H^2}{H_0^2} = \frac{\Omega_{m,0}}{a^3} = \Omega_{m,0}(1+z)^3. \tag{9.35}$$

[10] Not that there was any tungsten around back then to be melted.

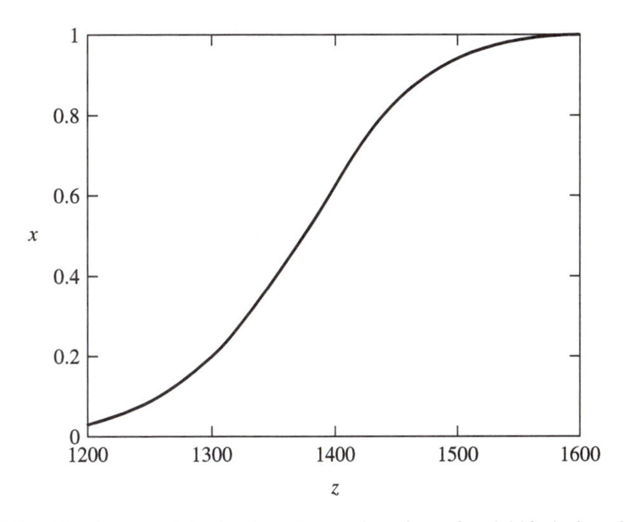

FIGURE 9.4 The fractional ionization X as a function of redshift during the epoch of recombination. A baryon-to-photon ratio of $\eta = 5.5 \times 10^{-10}$ is assumed.

Using $\Omega_{m,0} = 0.3$, the numerical value of the Hubble parameter during the epoch of recombination is

$$H(z) = 1.24 \times 10^{-18}\,\mathrm{s}^{-1}(1+z)^{3/2}. \tag{9.36}$$

The redshift of photon decoupling is found by setting $\Gamma = H$, or (combining equations (9.34) and (9.36)),

$$1 + z_{\mathrm{dec}} = \frac{43.0}{X(z_{\mathrm{dec}})^{2/3}}. \tag{9.37}$$

Using the value of $X(z)$ given by the Saha equation (shown in Figure 9.4), the redshift of photon decoupling is found to be $z_{\mathrm{dec}} = 1130$. In truth, the exact redshift of photon decoupling is somewhat smaller than this value. The Saha equation assumes that the reaction $\mathrm{H} + \gamma \rightleftharpoons p + e^-$ is in equilibrium. However, when Γ starts to drop below H, the photoionization reaction is no longer in equilibrium. As a consequence, at redshifts smaller than ~ 1200, the fractional ionization X is larger than would be predicted by the Saha equation, and the decoupling of photons is therefore delayed. Without going into the details of the nonequilibrium physics, let's content ourselves by saying, in round numbers, $z_{\mathrm{dec}} \approx 1100$, corresponding to a temperature $T_{\mathrm{dec}} \approx 3000\,\mathrm{K}$, when the age of the universe was $t_{\mathrm{dec}} \approx 350{,}000\,\mathrm{yr}$ in the Benchmark Model.

When we examine the CMB with our microwave antennas, the photons we collect have been traveling straight toward us since the last time they scattered from a free electron. During a brief time interval $t \to t + dt$, the probability that a photon undergoes a scattering is $dP = \Gamma(t)dt$, where $\Gamma(t)$ is the scattering rate at time t. Thus, if we detect a CMB photon at time t_0, the expected number of scatterings it has undergone since an earlier time t is

$$\tau(t) = \int_t^{t_0} \Gamma(t)dt. \tag{9.38}$$

The dimensionless number τ is the *optical depth*. The time t for which $\tau = 1$ is the *time of last scattering*, and represents the time that has elapsed since a typical CMB photon last scattered from a free electron. If we change the variable of integration in equation (9.38) from t to a, we find that

$$\tau(a) = \int_a^1 \Gamma(a)\frac{da}{\dot{a}} = \int_a^1 \frac{\Gamma(a)}{H(a)}\frac{da}{a}, \tag{9.39}$$

using the fact that $H = \dot{a}/a$. Alternatively, we can find the optical depth as a function of redshift by making the substitution $1 + z = 1/a$:

$$\tau(z) = \int_0^z \frac{\Gamma(z)}{H(z)}\frac{dz}{1+z} = 0.0035 \int_0^z X(z)(1+z)^{1/2}dz. \tag{9.40}$$

Here, we have made use of equations (9.34) and (9.36).[11] As it turns out, the last scattering of a typical CMB photon occurs after the photoionization reaction $\mathrm{H} + \gamma \rightleftharpoons p + e^-$ falls out of equilibrium, so the Saha equation doesn't strictly apply. To sufficient accuracy for our purposes, we can state that the redshift of last scattering was comparable to the redshift of photon decoupling: $z_{\mathrm{ls}} \approx z_{\mathrm{dec}} \approx 1100$. Not all the CMB photons underwent their last scattering simultaneously; the universe doesn't choreograph its microphysics that well. If we scoop up two photons from the CMB, one may have undergone its last scattering at $z = 1200$, while the other may have scattered more recently, at $z = 1000$. Thus, the "last scattering surface" is really more of a "last scattering layer"; just as we can see a little way into a fog bank here on Earth, we can see a little way into the "electron fog" that hides the early universe from our direct view.

The relevant times of various events around the time of recombination are shown in Table 9.1. For purposes of comparison, the table also contains the time of radiation-matter equality, emphasizing the fact that recombination, photon decoupling, and last scattering took place when the universe was matter-dominated. Note that all these times are approximate, and are dependent on the cosmologi-

TABLE 9.1 Events in the early universe

Event	Redshift	Temperature (K)	Time (megayears)
radiation-matter equality	3570	9730	0.047
recombination	1370	3740	0.24
photon decoupling	1100	3000	0.35
last scattering	1100	3000	0.35

[11]By the time the universe becomes Λ dominated, the free electron density has fallen to negligibly small levels, so using the Hubble parameter for a matter-dominated universe is a justifiable approximation in computing τ.

cal model you choose. (I have chosen the Benchmark Model in calculating these numbers.) When we look at the CMB, we are getting an intriguing glimpse of the universe as it was when it was only $1/40{,}000$ of its present age.

The epoch of photon decoupling marked an important change in the state of the universe. Before photon decoupling, there existed a single photon-baryon fluid, consisting of photons, electrons, and protons coupled together. Since the photons traveled at about the speed of light, kicking the electrons before them as they went, they tended to smooth out any density fluctuations in the photon-baryon fluid smaller than the horizon. After photon decoupling, however, the photon-baryon fluid became a pair of gases, one of photons and the other of neutral hydrogen. Although the two gases coexisted spatially, they were no longer coupled together. Thus, instead of being kicked to and fro by the photons, the hydrogen gas was free to collapse under its own self-gravity (and the added gravitational attraction of the dark matter). Thus, when we look at the Cosmic Microwave Background, we are looking backward in time to an important epoch in the history of the universe—the epoch when the baryons, free from the harassment of photons, were free to collapse gravitationally, and thus form galaxies, stars, planets, cosmologists, and the other dense knots of baryonic matter that make the universe so rich and strange today.

9.4 ■ TEMPERATURE FLUCTUATIONS

The dipole distortion of the Cosmic Microwave Background, shown in the top panel of Figure 9.2, results from the fact that the universe is not perfectly homogeneous today (at $z = 0$). Because we are gravitationally accelerated toward the nearest large lumps of matter, we see a Doppler shift in the radiation of the CMB. The distortions on a smaller angular scale, shown in the bottom panel of Figure 9.2, tell us that the universe was not perfectly homogeneous at the time of last scattering (at $z \approx 1100$). The angular size of the temperature fluctuations reflects in part the physical size of the density and velocity fluctuations at $z \approx 1100$. The COBE DMR experiment had limited angular resolution, and was only able to detect temperature fluctuations larger than $\delta\theta \approx 7°$. More recent experiments have provided higher angular resolution. For instance, MAXIMA (a balloon-borne experiment), DASI (an experiment located at the South Pole), and BOOMERANG (a balloon-borne experiment launched from Antarctica), all have provided maps of $\delta T/T$ down to scales of $\delta\theta \sim 10$ arcminutes.

The angular size $\delta\theta$ of a temperature fluctuation in the CMB is related to a physical size ℓ on the last scattering surface by the relation

$$d_A = \frac{\ell}{\delta\theta}, \tag{9.41}$$

where d_A is the angular-diameter distance to the last scattering surface. Since the last scattering surface is at a redshift $z_{\rm ls} = 1100 \gg 1$, a good approximation to

d_A is given by equation (7.41):

$$d_A \approx \frac{d_{\rm hor}(t_0)}{z_{\rm ls}}. \tag{9.42}$$

In the Benchmark Model, the current horizon distance is $d_{\rm hor}(t_0) \approx 14{,}000\,{\rm Mpc}$, so the angular-diameter distance to the surface of last scattering is

$$d_A \approx \frac{14{,}000\,{\rm Mpc}}{1100} \approx 13\,{\rm Mpc}. \tag{9.43}$$

Thus, fluctuations on the last scattering surface with an observed angular size $\delta\theta$ had a proper size

$$\begin{aligned} \ell = d_A(\delta\theta) &= 13\,{\rm Mpc}\left(\frac{\delta\theta}{1\,{\rm rad}}\right) \\ &= 0.22\,{\rm Mpc}\left(\frac{\delta\theta}{1^\circ}\right) \end{aligned} \tag{9.44}$$

at the time of last scattering. Thus, the fluctuations that gave rise to the fluctuations seen by COBE (with $\delta\theta > 7^\circ$) had a proper size $\ell > 1.6\,{\rm Mpc}$. However, the fluctuations at the time of last scattering were not gravitationally bound objects; they were expanding along with the universal Hubble expansion. Thus, the fluctuations seen by COBE correspond to physical scales of $\ell(1 + z_{\rm ls}) > 1700\,{\rm Mpc}$ today, much larger than the biggest superclusters. The higher-resolution experiments such as MAXIMA, DASI, and BOOMERANG see fluctuations corresponding to scales as small as $\ell \approx 0.04\,{\rm Mpc}$ at the time of last scattering, or $\ell(1 + z_{\rm ls}) \approx 40\,{\rm Mpc}$ today, about the size of today's superclusters.

Consider the density fluctuations $\delta T/T$ observed by a particular experiment. Figure 9.5, for instance, shows $\delta T/T$ as measured by COBE at low resolution over the entire sky, and as measured by BOOMERANG at higher resolution over part of the sky. Since $\delta T/T$ is defined on the surface of a sphere—the celestial sphere, in this case—it is useful to expand it in spherical harmonics:

$$\frac{\delta T}{T}(\theta, \phi) = \sum_{l=0}^{\infty} \sum_{m=-l}^{l} a_{lm} Y_{lm}(\theta, \phi), \tag{9.45}$$

where $Y_{lm}(\theta, \phi)$ are the usual spherical harmonic functions. What concerns cosmologists is not the exact pattern of hot spots and cold spots on the sky, but their statistical properties. The most important statistical property of $\delta T/T$ is the correlation function $C(\theta)$. Consider two points on the last scattering surface. Relative to an observer, they are in the directions $\hat{n}$ and $\hat{n}'$, and are separated by an angle θ given by the relation $\cos\theta = \hat{n}\cdot\hat{n}'$. To find the correlation function $C(\theta)$, multiply together the values of $\delta T/T$ at the two points, then average the product over all points separated by the angle θ:

$$C(\theta) = \left\langle \frac{\delta T}{T}(\hat{n}) \frac{\delta T}{T}(\hat{n}') \right\rangle_{\hat{n}\cdot\hat{n}'=\cos\theta}. \tag{9.46}$$

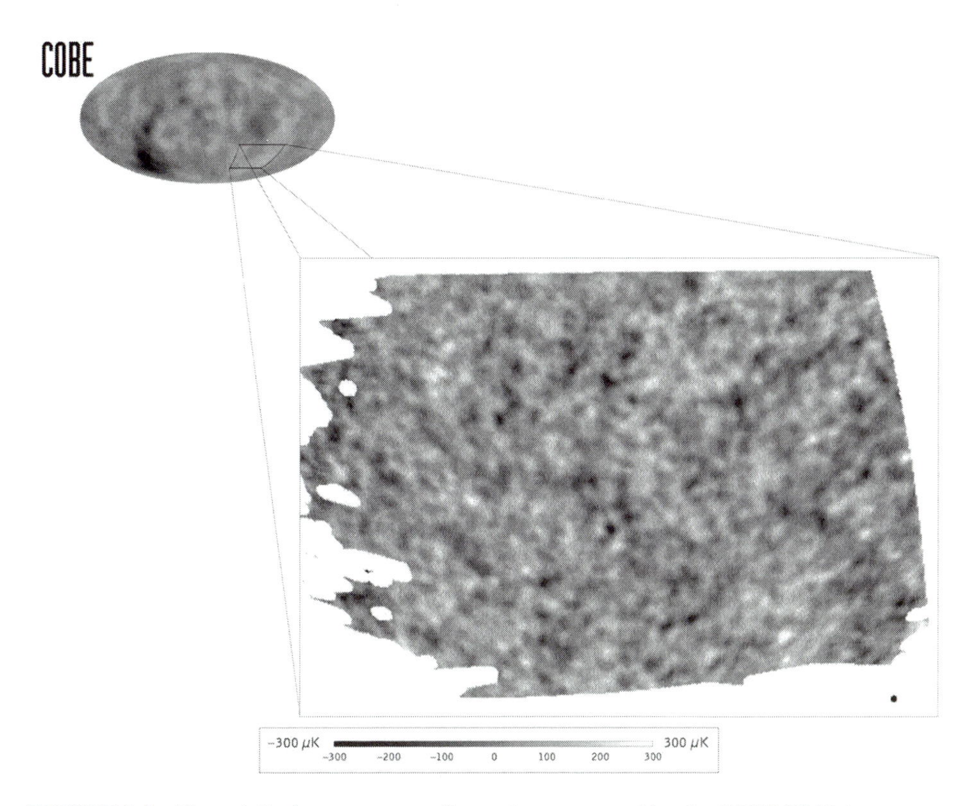

FIGURE 9.5 Upper left: the temperature fluctuations measured by the COBE DMR instrument, after the dipole and galaxy foreground are subtracted. Lower right: the temperature fluctuations measured by the BOOMERANG experiment over a region $\sim 20°$ across. The white spots are regions 300 microKelvin warmer than average; the black spots are regions 300 microKelvin cooler than average. (Courtesy of the BOOMERANG collaboration.)

If cosmologists knew the precise value of $C(\theta)$ for all angles from $\theta = 0$ to $\theta = 180°$, they would have a complete statistical description of the temperature fluctuations over all angular scales. Unfortunately, the CMB measurements that tell us about $C(\theta)$ contain information over only a limited range of angular scales.

The limited angular resolution of available observations is what makes the spherical harmonic expansion of $\delta T/T$, shown in equation (9.45), so useful. Using the expansion of $\delta T/T$ in spherical harmonics, the correlation function can be written in the form

$$C(\theta) = \frac{1}{4\pi} \sum_{l=0}^{\infty} (2l+1) C_l P_l(\cos\theta), \tag{9.47}$$

where P_l are the usual Legendre polynomials:

$$\begin{aligned} P_0(x) &= 1 \\ P_1(x) &= x \\ P_2(x) &= \frac{1}{2}(3x^2 - 1) \end{aligned} \tag{9.48}$$

and so forth. In this way, a measured correlation function $C(\theta)$ can be broken down into its multipole moments C_l. For a given experiment, the value of C_l will be nonzero for angular scales larger than the resolution of the experiment and smaller than the patch of sky examined. Generally speaking, a term C_l is a measure of temperature fluctuations on the angular scale $\theta \sim 180°/l$. Thus, the multipole l is interchangeable, for all practical purposes, with the angular scale θ. The $l = 0$ (monopole) term of the correlation function vanishes if we've defined the mean temperature correctly. The $l = 1$ (dipole) term results primarily from the Doppler shift due to our motion through space. The moments with $l \geq 2$ are of the most interest to cosmologists, since they tell us about the fluctuations present at the time of last scattering.

In presenting the results of CMB observations, it is customary to plot the function

$$\Delta_T \equiv \left(\frac{l(l+1)}{2\pi} C_l \right)^{1/2} \langle T \rangle \tag{9.49}$$

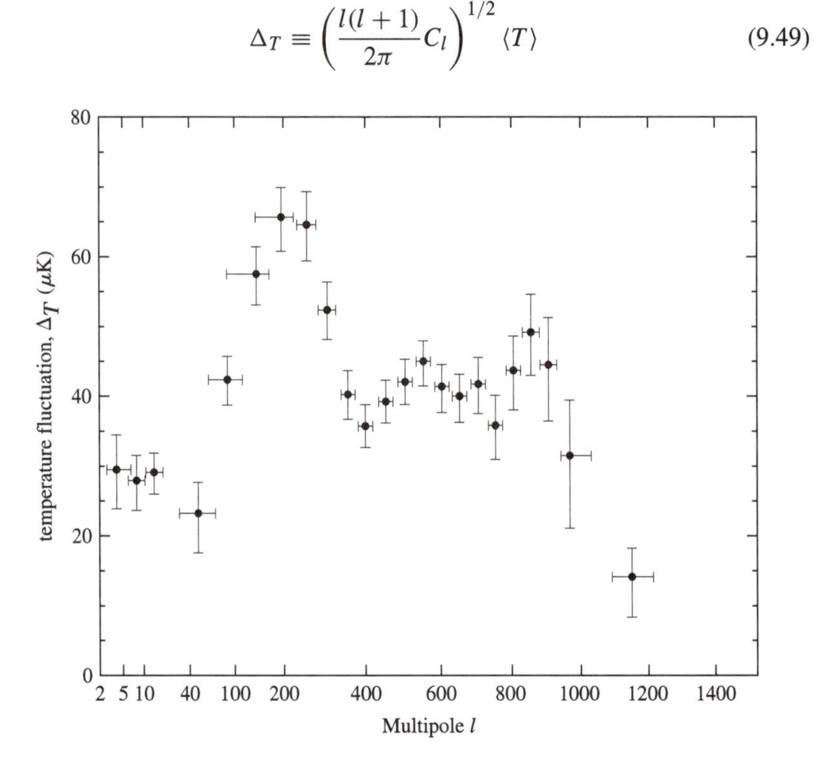

FIGURE 9.6 The anisotropy of the CMB temperature, Δ_T, expressed as a function of the multipole l.

since this function tells us the contribution per logarithmic interval in l to the total temperature fluctuation δT of the Cosmic Microwave Background. Figure 9.6, which combines data from a large number of CMB experiments, is a plot of Δ_T as a function of the logarithm of l. Note that the temperature fluctuation has a peak at $l \sim 200$, corresponding to an angular size of $\sim 1°$. The detailed shape of the Δ_T versus l curve, as shown in Figure 9.6, contains a wealth of information about the universe at the time of photon decoupling. In the next section we will examine, *very* briefly, the physics behind the temperature fluctuations, and how we can extract cosmological information from the temperature anisotropy of the Cosmic Microwave Background.

9.5 ■ WHAT CAUSES THE FLUCTUATIONS?

At the time of last scattering, a particularly interesting length scale, cosmologically speaking, is the Hubble distance,

$$c/H(z_{\rm ls}) \approx \frac{3.0 \times 10^8\,{\rm m\,s^{-1}}}{1.24 \times 10^{-18}\,{\rm s^{-1}}(1101)^{3/2}} \approx 6.6 \times 10^{21}\,{\rm m} \approx 0.2\,{\rm Mpc}, \quad (9.50)$$

where we have used equation (9.36) to compute the Hubble parameter at the redshift of last scattering, $z_{\rm ls} \approx 1100$. A patch of the last scattering surface with this physical size will have an angular size, as seen from Earth, of

$$\theta_H = \frac{c/H(z_{\rm ls})}{d_A} \approx \frac{0.2\,{\rm Mpc}}{13\,{\rm Mpc}} \approx 0.015\,{\rm rad} \approx 1°. \quad (9.51)$$

It is no coincidence that the peak in the Δ_T versus l curve (Figure 9.6) occurs at an angular scale $\theta \sim \theta_H$. The origin of temperature fluctuations with $\theta > \theta_H$ ($l < 180$) is different from those with $\theta < \theta_H$ ($l > 180$).

Consider first the large-scale fluctuations—those with angular size $\theta > \theta_H$. These temperature fluctuations arise from the gravitational effect of primordial density fluctuations in the distribution of nonbaryonic dark matter. The density of nonbaryonic dark matter at the time of last scattering was

$$\varepsilon_{\rm dm}(z_{\rm ls}) = \Omega_{\rm dm,0}\varepsilon_{c,0}(1 + z_{\rm ls})^3 \quad (9.52)$$

since the energy density of matter is $\propto a^{-3} \propto (1+z)^3$. Plugging in the appropriate numbers, we find that

$$\varepsilon_{\rm dm}(z_{\rm ls}) \approx (0.26)(5200\,{\rm MeV\,m^{-3}})(1101)^3 \approx 1.8 \times 10^{12}\,{\rm MeV\,m^{-3}}, \quad (9.53)$$

equivalent to a mass density of $\sim 3 \times 10^{-18}\,{\rm kg\,m^{-3}}$. The density of baryonic matter at the time of last scattering was

$$\varepsilon_{\rm bary}(z_{\rm ls}) = \Omega_{\rm bary,0}\epsilon_{c,0}(1 + z_{\rm ls})^3 \approx 2.8 \times 10^{11}\,{\rm MeV\,m^{-3}}. \quad (9.54)$$

The density of photons at the time of last scattering, since $\varepsilon_\gamma \propto a^{-4} \propto (1+z)^4$, was

$$\varepsilon_\gamma(z_{\rm ls}) = \Omega_{\gamma,0}\epsilon_{c,0}(1+z_{\rm ls})^4 \approx 3.8 \times 10^{11}\,{\rm MeV\,m^{-3}}. \tag{9.55}$$

Thus, at the time of last scattering, $\varepsilon_{\rm dm} > \varepsilon_\gamma > \varepsilon_{\rm bary}$, with dark matter, photons, and baryons having energy densities in roughly the ratio 6.4 : 1.4 : 1. The nonbaryonic dark matter dominated the energy density ϵ, and hence the gravitational potential, of the universe at the time of last scattering.

Suppose that the density of the nonbaryonic dark matter at the time of last scattering was not perfectly homogeneous, but varied as a function of position. Then we could write the energy density of the dark matter as

$$\varepsilon(\vec{r}) = \bar{\varepsilon} + \delta\varepsilon(\vec{r}), \tag{9.56}$$

where $\bar{\varepsilon}$ is the spatially averaged energy density of the nonbaryonic dark matter, and $\delta\varepsilon$ is the local deviation from the mean. In the Newtonian approximation, the spatially varying component of the energy density, $\delta\varepsilon$, gives rise to a spatially varying gravitational potential $\delta\Phi$. The link between $\delta\varepsilon$ and $\delta\Phi$ is Poisson's equation:

$$\nabla^2(\delta\Phi) = \frac{4\pi G}{c^2}\delta\varepsilon. \tag{9.57}$$

Unless the distribution of dark matter were perfectly smooth at the time of last scattering, the fluctuations in its density would necessarily have given rise to fluctuations in the gravitational potential.

Consider the fate of a CMB photon that happens to be at a local minimum of the potential at the time of last scattering. (Minima in the gravitational potential are known colloquially as "potential wells.") In climbing out of the potential well, it loses energy, and consequently is redshifted. Conversely, a photon that happens to be at a potential maximum when the universe became transparent gains energy as it falls down the "potential hill," and thus is blueshifted. The cool (redshifted) spots on the COBE temperature map correspond to minima in $\delta\Phi$ at the time of last scattering; the hot (blueshifted) spots correspond to maxima in $\delta\Phi$. A detailed general relativistic calculation, first performed by Sachs and Wolfe in 1967, tells us that

$$\frac{\delta T}{T} = \frac{1}{3}\frac{\delta\Phi}{c^2}. \tag{9.58}$$

Thus, the temperature fluctuations on large angular scales ($\theta > \theta_H \approx 1°$) give us a map of the potential fluctuations $\delta\Phi$ present at the time of last scattering. The creation of temperature fluctuations by variations in the gravitational potential is known as the *Sachs–Wolfe effect*, in tribute to the work of Sachs and Wolfe.

On smaller scales ($\theta < \theta_H$), the origin of the temperature fluctuations in the CMB is complicated by the behavior of the photons and baryons. Consider the situation immediately prior to photon decoupling. The photons, electrons, and protons together make a single photon-baryon fluid, whose energy density is only about a third that of the dark matter. Thus, the photon-baryon fluid moves primarily under the gravitational influence of the dark matter, rather than under its own self-gravity. The equation-of-state parameter $w_{\rm pb}$ of the photon-baryon fluid is intermediate between the value $w = \frac{1}{3}$ expected for a gas containing only photons and the value $w = 0$ expected for a gas containing only cold baryons and electrons. If the photon-baryon fluid finds itself in a potential well of the dark matter, it will fall to the center of the well.[12] As the photon-baryon fluid is compressed by gravity, however, its pressure starts to rise. Eventually, the pressure is sufficient to cause the fluid to expand outward. As the expansion continues, the pressure drops until gravity causes the photon-baryon fluid to fall inward again. The cycle of compression and expansion continues thus until the time of photon decoupling. The inward and outward oscillations of the photon-baryon fluid are called *acoustic oscillations*, since they represent a type of standing sound wave in the photon-baryon fluid.

If the photon-baryon fluid within a potential well is at maximum compression at the time of photon decoupling, its density will be higher than average, and since $T \propto \varepsilon^{1/4}$, the liberated photons will be hotter than average. Conversely, if the photon-baryon fluid within a potential well is at maximum expansion at the time of decoupling, the liberated photons will be slightly cooler than average. If the photon-baryon fluid is in the process of expanding or contracting at the time of decoupling, the Doppler effect will cause the liberated photons to be cooler or hotter than average, depending on whether the photon-baryon fluid was moving away from our location or toward it at the time of photon decoupling. Computing the exact shape of the Δ_T versus l curve expected in a particular model universe is a rather complicated chore. Generally speaking, however, the highest peak in the Δ_T curve (at $l \sim 200$ or $\theta \sim 1^\circ$ in Figure 9.6) represents the potential wells within which the photon-baryon fluid had just reached maximum compression at the time of last scattering. These potential wells had proper sizes $\sim c/H(z_{\rm ls})$ at the time of last scattering, and hence have angular sizes of $\sim \theta_H$ as seen by us at the present day.

The location and amplitude of the highest peak in Figure 9.6 is a very useful cosmological diagnostic. As it turns out, the angle $\theta \approx 1^\circ$ at which it's located is dependent on the spatial curvature of the universe. In a negatively curved universe ($\kappa = -1$), the angular size θ of an object of known proper size at a known redshift is *smaller* than it is in a positively curved universe ($\kappa = +1$). If the universe were negatively curved, the peak in Δ_T would be seen at an angle $\theta < 1^\circ$, or $l > 180$; if the universe were positively curved, the peak would be seen at an angle $\theta > 1^\circ$, or $l < 180$. The observed position of the peak is consistent with $\kappa = 0$, or $\Omega_0 = 1$. Figure 9.7 shows the values of $\Omega_{m,0}$ and $\Omega_{\Lambda,0}$ permitted by the present

[12]Note that if the size of the well is larger than $c/H(z_{\rm ls})$, the photon-baryon fluid, which travels at a speed $< c$, will not have time to fall to the center by the time of last scattering $t_{\rm ls} \sim 1/H(z_{\rm ls})$. This is why the motions of the photons and baryons are irrelevant on scales $\theta > \theta_H$, and why the temperature fluctuations on these large scales are dictated purely by the distribution of dark matter.

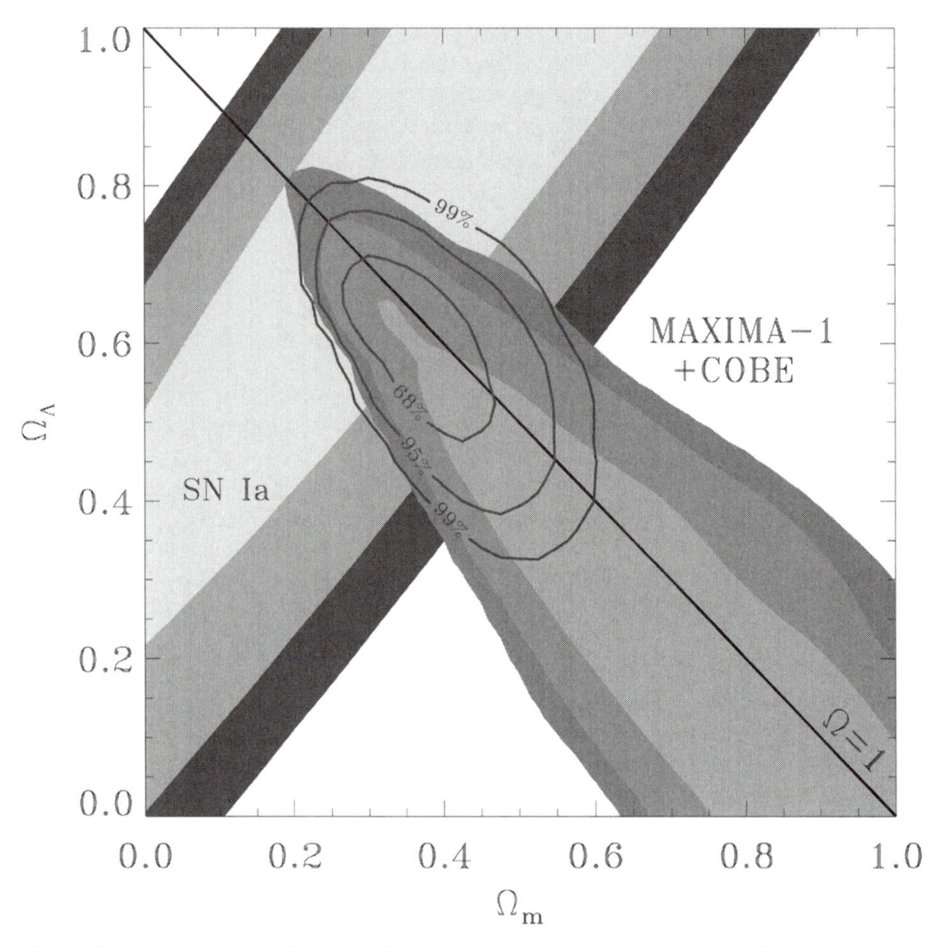

FIGURE 9.7 The values of $\Omega_{m,0}$ and $\Omega_{\Lambda,0}$ permitted by the MAXIMA and COBE DMR data. The light gray area near the $\Omega = 1$ line shows the best-fitting values of $\Omega_{m,0}$ and $\Omega_{\Lambda,0}$. The SN Ia data (as seen earlier in Figure 7.6) are also shown. The curves marked 68%, 95%, and 99% give the best fits, at the designated confidence levels, for the *combined* CMB and SN Ia data (from Stompor et al., 2001, ApJ, 561, L7).

CMB data. Note that the shaded areas that show the best fit to the CMB data are roughly parallel to the $\Omega = 1$ line (representing $\kappa = 0$ universes) and are roughly perpendicular to the region permitted by the type Ia supernova results. The oval curve marked "68%" in Figure 9.7 represents the region consistent with both the CMB results and the supernova results. A spatially flat universe, with $\Omega_{m,0} \approx 0.3$ and $\Omega_{\Lambda,0} \approx 0.7$, agrees with the CMB results, the supernova results, and the computed density of matter in clusters (as discussed in Chapter 8). This happy concurrence is the basis for the "Benchmark Model" we have been using in this book.

The angular size corresponding to the highest peak in the Δ_T versus l curve gives useful information about the density parameter Ω and the curvature κ. The amplitude of the peak is dependent on the sound speed of the photon-baryon fluid prior to photon decoupling. Since the sound speed of the photon-baryon fluid is

$$c_s = \sqrt{w_{\rm pb}}\, c, \tag{9.59}$$

and the equation-of-state parameter $w_{\rm pb}$ is in turn dependent on the baryon-to-photon ratio, the amplitude of the peak is a useful diagnostic of the baryon density of the universe. Detailed analysis of the Δ_T curve, with the currently available data, yields

$$\Omega_{\rm bary,0} = 0.04 \pm 0.02, \tag{9.60}$$

assuming a Hubble constant of $H_0 = 70\,{\rm km\,s^{-1}\,Mpc^{-1}}$. As we'll see in the next chapter, this baryon density is consistent with that found from the entirely different arguments of primordial nucleosynthesis.

SUGGESTED READING

Full references are given in the Annotated Bibliography on page 235.

Coles (1999), Part I, ch. 5: An insightful overview of the origin of the temperature anisotropies in the CMB.

Liddle (1999), ch. 9: A simplified overview of photon decoupling and the origin of the CMB.

Rich (2001), ch. 7.9: A discussion of the anisotropies in the CMB, and their use as diagnostics of cosmological parameters.

PROBLEMS

9.1. The purpose of this problem is to determine how the uncertainty in the value of the baryon-to-photon ratio, η, affects the recombination temperature in the early universe. Plot the fractional ionization X as a function of temperature, in the range $3000\,{\rm K} < T < 4500\,{\rm K}$; first make the plot assuming $\eta = 4 \times 10^{-10}$, then assuming $\eta = 8 \times 10^{-10}$. How much does this change in η affect the computed value of the recombination temperature $T_{\rm rec}$, if we define $T_{\rm rec}$ as the temperature at which $X = \frac{1}{2}$?

9.2. Suppose the temperature T of a blackbody distribution is such that $kT \ll Q$, where $Q = 13.6\,{\rm eV}$ is the ionization energy of hydrogen. What fraction f of the blackbody photons are energetic enough to ionize hydrogen? If $T = T_{\rm rec} = 3740\,{\rm K}$, what is the numerical value of f?

9.3. Imagine that at the time of recombination, the baryonic portion of the universe consisted entirely of ^{4}He (that is, helium with two protons and two neutrons in its nucleus). The ionization energy of helium (that is, the energy required to convert neutral He to He^+) is $Q_{He} = 24.6\,eV$. At what temperature would the fractional ionization of the helium be $X = \frac{1}{2}$? Assume that $\eta = 5.5 \times 10^{-10}$ and that the number density of He^{++} is negligibly small. [The relevant statistical weight factor for the ionization of helium is $g_{He}/(g_e g_{He^+}) = \frac{1}{4}$.]

9.4. What is the proper distance d_p to the surface of last scattering? What is the luminosity distance d_L to the surface of last scattering? Assume that the Benchmark Model is correct, and that the redshift of the last scattering surface is $z_{ls} = 1100$.

9.5. We know from observations that the intergalactic medium is currently ionized. Thus, at some time between t_{rec} and t_0, the intergalactic medium must have been *reionized*. The fact that we can see small fluctuations in the CMB places limits on how early the reionization took place. Assume that the baryonic component of the universe instantaneously became completely reionized at some time t_*. For what value of t_* does the optical depth of reionized material,

$$\tau = \int_{t_*}^{t_0} \Gamma(t)dt = \int_{t_*}^{t_0} n_e(t)\sigma_e c dt, \tag{9.61}$$

equal one? For simplicity, assume that the universe is spatially flat and matter-dominated, and that the baryonic component of the universe is pure hydrogen. To what redshift z_* does this value of t_* correspond?

CHAPTER 10
Nucleosynthesis and the Early Universe

The Cosmic Microwave Background tells us a great deal about the state of the universe at the time of last scattering ($t_{\rm ls} \approx 0.35\,\text{Myr}$). However, the opacity of the early universe prevents us from directly seeing what the universe was like at $t < t_{\rm ls}$. Photons are the "messenger boys" of astronomy; although some information is carried by cosmic rays and by neutrinos, most of what we know about the universe beyond our solar system has come in the form of photons. Looking at the last scattering surface is like looking at the surface of a cloud, or the surface of the Sun; our curiosity is piqued, and we wish to find out what conditions are like in the opaque regions so tantalizingly hidden from our direct view.

Theoretically, many properties of the early universe should be quite simple. For instance, when radiation is strongly dominant over matter, at scale factors $a \ll a_{rm} \approx 2.8 \times 10^{-4}$, or times $t \ll t_{rm} \approx 47{,}000\,\text{yr}$, the expansion of the universe has the simple power-law form $a(t) \propto t^{1/2}$. The temperature of the blackbody photons in the early universe, which decreases as $T \propto a^{-1}$ as the universe expands, is given by the convenient relation

$$T(t) \approx 10^{10}\,\text{K}\left(\frac{t}{1\,\text{s}}\right)^{-1/2}, \tag{10.1}$$

or equivalently

$$kT(t) \approx 1\,\text{MeV}\left(\frac{t}{1\,\text{s}}\right)^{-1/2}. \tag{10.2}$$

Thus the mean energy per photon was

$$E_{\rm mean}(t) \approx 2.7kT(t) \approx 3\,\text{MeV}\left(\frac{t}{1\,\text{s}}\right)^{-1/2}. \tag{10.3}$$

The Fermi National Accelerator Laboratory in Batavia, Illinois, is justly proud of its "Tevatron," designed to accelerate protons to an energy of 1 TeV; that's 10^{12} eV, or a thousand times the rest energy of a proton. Well, when the universe was one picosecond old ($t = 10^{-12}$ s), the entire universe was a Tevatron. Thus, the early universe is referred to as "the poor man's particle accelerator," since it provided particles of very high energy without running up an enormous electricity bill or having Congress threaten to cut off its funding.

10.1 ■ NUCLEAR PHYSICS AND COSMOLOGY

As the universe has expanded and cooled, the mean energy per photon has dropped from $E_{\rm mean}(t_P) \sim E_P \sim 10^{28}\,{\rm eV}$ at the Planck time to $E_{\rm mean}(t_0) \approx 6 \times 10^{-4}\,{\rm eV}$ at the present day. Thus, by studying the universe as it expands, we sample over 31 orders of magnitude in particle energy. Within this wide energy range, some energies are of more interest than others to physicists. For instance, to physicists studying recombination and photoionization, the most interesting energy scale is the ionization energy of an atom. The ionization energy of hydrogen is $Q = 13.6\,{\rm eV}$, as we have already noted. The ionization energies of other elements (that is, the energy required to remove the most loosely bound electron in the neutral atom) are roughly comparable. Thus, atomic physicists, when considering the ionization of atoms, typically deal with energies of $\sim 10\,{\rm eV}$, in round numbers.

Nuclear physicists are concerned not with ionization and recombination (removing or adding electrons to an atom), but with the much higher energy processes of fission and fusion (splitting or merging atomic nuclei). An atomic nucleus contains Z protons and N neutrons, where $Z \geq 1$ and $N \geq 0$. Protons and neutrons are collectively called *nucleons*. The total number of nucleons within an atomic nucleus is called the *mass number*, and is given by the formula $A = Z + N$. The proton number Z of a nucleus determines the atomic element to which that nucleus belongs. For instance, hydrogen (H) nuclei all have $Z = 1$, helium (He) nuclei have $Z = 2$, lithium (Li) nuclei have $Z = 3$, beryllium (Be) nuclei have $Z = 4$, and so on, through the complete periodic table. Although all atoms of a given element have the same number of protons in their nuclei, different isotopes of an element can have different numbers of neutrons. A particular isotope of an element is designated by prefixing the mass number A to the symbol for that element. For instance, a standard hydrogen nucleus, with one proton and no neutrons, is symbolized as ^{1}H. (Since an ordinary hydrogen nucleus is nothing more than a proton, we may also write p in place of ^{1}H when considering nuclear reactions.) Heavy hydrogen, or *deuterium*, contains one proton and one neutron, and is symbolized as ^{2}H. (Since deuterium is very frequently mentioned in the context of nuclear fusion, it has its own special symbol, D.) Ordinary helium contains two protons and two neutrons, and is symbolized as ^{4}He.

The binding energy B of a nucleus is the energy required to pull it apart into its component protons and neutrons. Equivalently, it is the energy released when a nucleus is fused together from individual protons and neutrons. For instance, when a neutron and a proton are bound together to form a deuterium nucleus, an energy of $B_{\rm D} = 2.22\,{\rm MeV}$ is released:

$$p + n \rightleftharpoons {\rm D} + 2.22\,{\rm MeV}. \tag{10.4}$$

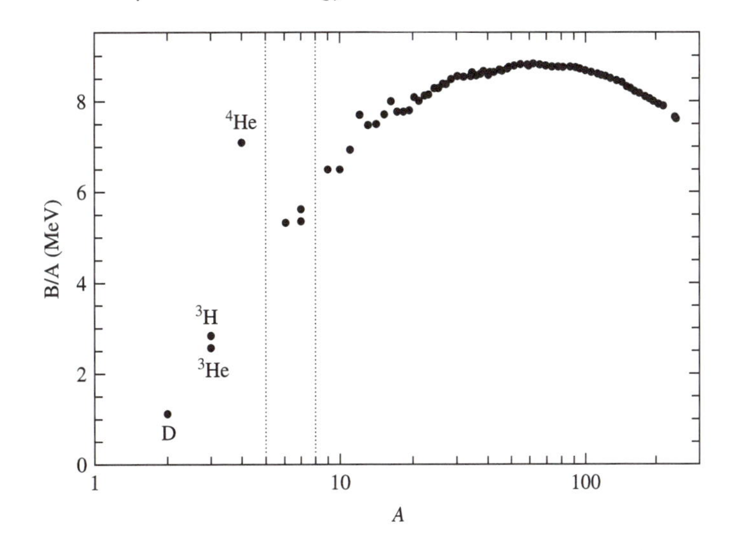

FIGURE 10.1 The binding energy per nucleon (B/A) as a function of the number of nucleons (protons and neutrons) in an atomic nucleus. Note the absence of nuclei at $A = 5$ and $A = 8$.

The deuterium nucleus is not very tightly bound, compared to other atomic nuclei. Figure 10.1 plots the binding energy per nucleon (B/A) for atomic nuclei with different mass numbers. Note that ^{4}He, with a total binding energy of $B = 28.30\,\text{MeV}$, and a binding energy per nucleon of $B/A = 7.07\,\text{MeV}$, is relatively tightly bound, compared to other light nuclei (that is, nuclei with $A \leq 10$). The most tightly bound nuclei are those of ^{56}Fe and ^{62}Ni, which both have $B/A \approx 8.8\,\text{MeV}$. Thus, nuclei more massive than iron or nickel can release energy by fission—splitting into lighter nuclei. Nuclei less massive than iron or nickel can release energy by fusion—merging into heavier nuclei.

Thus, just as studies of ionization and recombination deal with an energy scale of $\sim 10\,\text{eV}$ (a typical ionization energy), so studies of nuclear fusion and fission deal with an energy scale of $\sim 8\,\text{MeV}$ (a typical binding energy per nucleon). Moreover, just as electrons and protons combined to form neutral hydrogen atoms when the temperature dropped sufficiently far below the ionization energy of hydrogen ($Q = 13.6\,\text{eV}$), so protons and neutrons must have fused to form deuterium when the temperature dropped sufficiently far below the binding energy of deuterium ($B_\text{D} = 2.22\,\text{MeV}$). The epoch of recombination must have been preceded by an epoch of nuclear fusion, commonly called the epoch of Big Bang Nucleosynthesis (BBN). Nucleosynthesis in the early universe starts by the fusion of neutrons and protons to form deuterium, then proceeds to form heavier nuclei by successive acts of fusion. Since the binding energy of deuterium is larger than the ionization energy of hydrogen by a factor $B_\text{D}/Q = 1.6 \times 10^5$, we would expect, as a rough estimate, the synthesis of deuterium to occur at a temperature 1.6×10^5 times higher than the recombination temperature $T_\text{rec} = 3740\,\text{K}$. That is, deuterium synthesis occurred at a temperature $T_{\rm nuc} \approx 1.6 \times 10^5 (3740\,{\rm K}) \approx 6 \times 10^8\,{\rm K}$, corresponding to a time $t_{\rm nuc} \approx 300\,{\rm s}$. This estimate, as we'll see when we do the detailed calculations, gives a temperature slightly too low, but it certainly gives the right order of magnitude. As indicated in the title of Steven Weinberg's classic book, *The First Three Minutes*, the entire saga of Big Bang Nucleosynthesis takes place when the universe is only a few minutes old.

One thing we can say about Big Bang Nucleosynthesis, after taking a look at the present-day universe, is that it was shockingly inefficient. From an energy viewpoint, the preferred universe would be one in which the baryonic matter consisted of an iron-nickel alloy. Obviously, we do not live in such a universe. Currently, three-fourths of the baryonic component (by mass) is still in the form of unbound protons, or ^{1}H. Moreover, when we look for nuclei heavier than ^{1}H, we find that they are primarily ^{4}He, a relatively lightweight nucleus compared to ^{56}Fe and ^{62}Ni. The primordial helium fraction of the universe (that is, the helium fraction before nucleosynthesis begins in stars) is usually expressed as the dimensionless number

$$Y_p \equiv \frac{\rho(^4{\rm He})}{\rho_{\rm bary}}. \tag{10.5}$$

That is, Y_p is the mass density of ^{4}He divided by the mass density of all the baryonic matter. The Sun's atmosphere has a helium fraction (by mass) of $Y = 0.28$. However, the Sun is made of recycled interstellar gas, which was contaminated by helium formed in earlier generations of stars. When we look at astronomical objects of different sorts, we find a minimum value of $Y = 0.24$. That is, baryonic objects such as stars and gas clouds are all at least 24% helium.[1]

10.2 ■ NEUTRONS AND PROTONS

The basic building blocks for nucleosynthesis are neutrons and protons. The rest energy of a neutron is greater than that of a proton by a factor

$$Q_n = m_n c^2 - m_p c^2 = 1.29\,{\rm MeV}. \tag{10.6}$$

A free neutron is unstable, decaying via the reaction

$$n \rightarrow p + e^- + \bar{\nu}_e. \tag{10.7}$$

The *decay time* for a free neutron is $\tau_n = 890\,{\rm s}$. That is, if you start with a population of free neutrons, after a time t, a fraction $f = \exp(-t/\tau_n)$ will remain.[2] Since the energy Q_n released by the decay of a neutron into a proton is greater

[1] Condensed objects that have undergone chemical or physical fractionation can be much lower in helium than this value. For instance, your helium fraction is $\ll 24\%$.

[2] The half-life, the time it takes for half the neutrons to decay, is related to the decay time by the relation $t_{1/2} = \tau_n \ln 2 = 617\,{\rm s}$.

than the rest energy of an electron ($m_e c^2 = 0.51$ MeV), the remainder of the energy is carried away by the kinetic energy of the electron and the energy of the electron anti-neutrino. With a decay time of only fifteen minutes, the existence of a free neutron is as fleeting as fame; once the universe was several hours old, it contained essentially no free neutrons. However, a neutron bound into a stable atomic nucleus is preserved against decay. Neutrons are still around today because they've been tied up in deuterium, helium, and other atoms.

Let's consider the state of the universe when its age was $t = 0.1$ s. At that time, the temperature was $T \approx 3 \times 10^{10}$ K, and the mean energy per photon was $E_{\rm mean} \approx 10$ MeV. This energy is much greater than the rest energy of a electron or positron, so there were positrons as well as electrons present at $t = 0.1$ s, created by pair production:

$$\gamma + \gamma \rightleftharpoons e^- + e^+. \tag{10.8}$$

At $t = 0.1$ s, neutrons and protons were in equilibrium with each other, via the interactions

$$n + \nu_e \rightleftharpoons p + e^- \tag{10.9}$$

and

$$n + e^+ \rightleftharpoons p + \bar{\nu}_e. \tag{10.10}$$

As long as neutrons and protons are kept in equilibrium by the reactions shown in equations (10.9) and (10.10), their number density is given by the Maxwell–Boltzmann equation, as discussed in section 9.3. The number density of neutrons is then

$$n_n = g_n \left(\frac{m_n kT}{2\pi\hbar^2}\right)^{3/2} \exp\left(-\frac{m_n c^2}{kT}\right) \tag{10.11}$$

and the number density of protons is

$$n_p = g_p \left(\frac{m_p kT}{2\pi\hbar^2}\right)^{3/2} \exp\left(-\frac{m_p c^2}{kT}\right). \tag{10.12}$$

Since the statistical weights of protons and neutrons are equal, with $g_p = g_n = 2$, the neutron-to-proton ratio, from equations (10.11) and (10.12), is

$$\frac{n_n}{n_p} = \left(\frac{m_n}{m_p}\right)^{3/2} \exp\left(-\frac{(m_n - m_p)c^2}{kT}\right). \tag{10.13}$$

The above equation can be simplified. First, $(m_n/m_p)^{3/2} = 1.002$; there will be no great loss in accuracy if we set this factor equal to one. Second, the difference in rest energy of the neutron and proton is $(m_n - m_p)c^2 = Q_n = 1.29$ MeV.

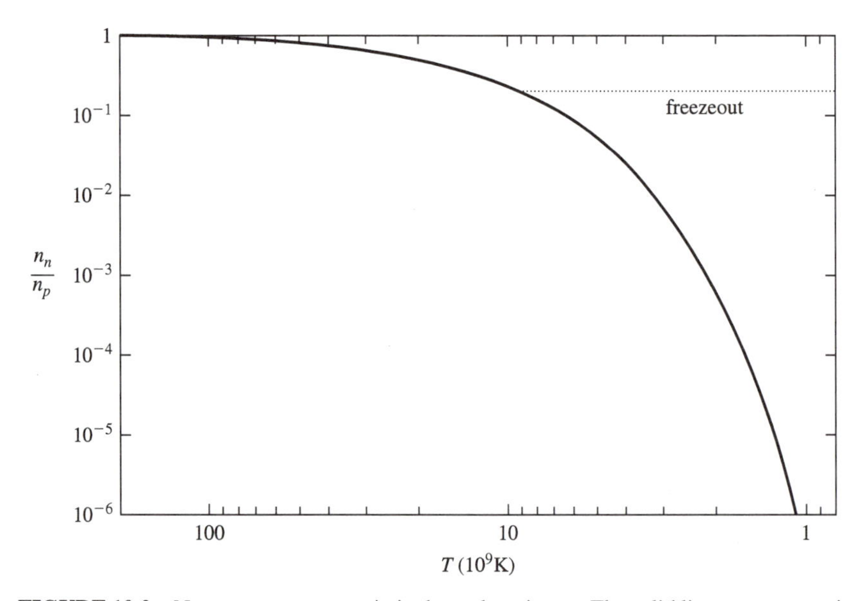

FIGURE 10.2 Neutron-to-proton ratio in the early universe. The solid line assumes equilibrium; the dotted line gives the value after freezeout.

Thus, the equilibrium neutron-to-proton ratio has the particularly simple form

$$\frac{n_n}{n_p} = \exp\left(-\frac{Q_n}{kT}\right), \tag{10.14}$$

illustrated as the solid line in Figure 10.2. At temperatures $kT \gg Q_n = 1.29\,\mathrm{MeV}$, corresponding to $T \gg 1.5 \times 10^{10}\,\mathrm{K}$ and $t \ll 1\,\mathrm{s}$, the number of neutrons is nearly equal to the number of protons. However, as the temperature starts to drop below $1.5 \times 10^{10}\,\mathrm{K}$, protons begin to be strongly favored, and the neutron-to-proton ratio plummets exponentially.

If the neutrons and protons remained in equilibrium, then by the time the universe was six minutes old, there would be only one neutron for every million protons. However, neutrons and protons do not remain in equilibrium for nearly that long. The interactions that mediate between neutrons and protons in the early universe, shown in equations (10.9) and (10.10), involve the interaction of a baryon with a neutrino (or anti-neutrino). Neutrinos interact with baryons via the weak nuclear force. The cross-sections for weak interactions have the temperature dependence $\sigma_w \propto T^2$; at the temperatures we are considering, the cross-sections are small. A typical cross-section for the interaction of a neutrino with any other particle via the weak nuclear force is

$$\sigma_w \sim 10^{-47}\,\mathrm{m}^2 \left(\frac{kT}{1\,\mathrm{MeV}}\right)^2. \tag{10.15}$$

(Compare this to the Thomson cross-section for the interaction of electrons via the electromagnetic force: $\sigma_e = 6.65 \times 10^{-29}\,\text{m}^2$.) In the radiation-dominated universe, the temperature falls at the rate $T \propto a(t)^{-1} \propto t^{-1/2}$, and thus the cross-sections for weak interactions diminish at the rate $\sigma_w \propto t^{-1}$. The number density of neutrinos falls at the rate $n_\nu \propto a(t)^{-3} \propto t^{-3/2}$, and hence the rate Γ with which neutrons and protons interact with neutrinos via the weak force falls rapidly:

$$\Gamma = n_\nu c \sigma_w \propto t^{-5/2}. \tag{10.16}$$

Meanwhile, the Hubble parameter is only decreasing at the rate $H \propto t^{-1}$. When $\Gamma \approx H$, the neutrinos decouple from the neutrons and protons, and the ratio of neutrons to protons is "frozen" (at least until the neutrons start to decay, at times $t \sim \tau_n$). An exact calculation of the temperature T_{freeze} at which $\Gamma = H$ requires a knowledge of the exact cross-section of the proton and neutron for weak interactions. Using the best available laboratory information, the "freezeout temperature" turns out to be $kT_{\text{freeze}} = 0.8\,\text{MeV}$, or $T_{\text{freeze}} = 9 \times 10^9\,\text{K}$. The universe reaches this temperature when its age is $t_{\text{freeze}} \sim 1\,\text{s}$. The neutron-to-proton ratio, once the temperature drops below T_{freeze}, is frozen at the value

$$\frac{n_n}{n_p} = \exp\left(-\frac{Q_n}{kT_{\text{freeze}}}\right) \approx \exp\left(-\frac{1.29\,\text{MeV}}{0.8\,\text{MeV}}\right) \approx 0.2. \tag{10.17}$$

At times $t_{\text{freeze}} < t \ll \tau_n$, there was one neutron for every five protons in the universe.

The scarcity of neutrons relative to protons explains why Big Bang Nucleosynthesis was so incomplete, leaving three-fourths of the baryons in the form of unfused protons. A neutron will fuse with a proton much more readily than a proton will fuse with another proton. When two protons fuse to form a deuterium nucleus, a positron must be emitted (to conserve charge); this means that an electron neutrino must also be emitted (to conserve electron quantum number). The proton-proton fusion reaction can be written as

$$p + p \rightarrow \text{D} + e^+ + \nu_e. \tag{10.18}$$

The involvement of a neutrino in this reaction tells us that it involves the weak nuclear force, and thus has a minuscule cross-section, of order σ_w. By contrast, the neutron-proton fusion reaction is

$$p + n \rightleftharpoons \text{D} + \gamma. \tag{10.19}$$

No neutrinos are involved; this is a strong interaction (one involving the strong nuclear force). The cross-section for interactions involving the strong nuclear force are much larger than for those involving the weak nuclear force. The rate of proton-proton fusion is much slower than the rate of neutron-proton fusion, for two reasons. First, the cross-section for proton-proton fusion, since it is a weak interaction, is minuscule compared to the cross-section for neutron-proton fusion. Second, since protons are all positively charged, they must surmount the Coulomb barrier between them in order to fuse.

It's possible, of course, to coax two protons into fusing with each other. It's happening in the Sun, for instance, even as you read this sentence. However, fusion in the Sun is a very slow process. If you pick out any particular proton in the Sun's core, it has only one chance in ten billion of undergoing fusion during the next year. Only exceptionally fast protons have any chance of undergoing fusion, and even those high-speed protons have only a tiny probability of quantum tunneling through the Coulomb barrier of another proton and fusing with it. The core of the Sun, though, is a stable environment; it's in hydrostatic equilibrium, and its temperature and density change only slowly with time. In the early universe, by strong contrast, the temperature drops as $T \propto t^{-1/2}$ and the density of baryons drops as $n_{\rm bary} \propto t^{-3/2}$. Big Bang Nucleosynthesis is a race against time. After less than an hour, the temperature and density have dropped too low for fusion to occur.

For the sake of completeness, note also that the rate of neutron-neutron fusion in the early universe is negligibly small compared to the rate of neutron-proton fusion. The reaction governing neutron-neutron fusion is

$$n + n \rightarrow {\rm D} + e^- + \bar{\nu}_e. \tag{10.20}$$

Again, the presence of a neutrino (an electron anti-neutrino, in this case) tells us this is an interaction involving the weak nuclear force. Thus, although there is no Coulomb barrier between neutrons, the neutron-neutron fusion rate is tiny. In part, this is because of the scarcity of neutrons relative to protons, but mainly it is because of the small cross-section for neutron-neutron fusion.

Given the alacrity of neutron-proton fusion when compared to the leisurely rate of proton-proton and neutron-neutron fusion, we can state, as a lowest order approximation, that BBN proceeds until every free neutron is bonded into an atomic nucleus, with the leftover protons remaining solitary. In this approximation, we can compute the maximum possible value of Y_p, the fraction of the baryon mass in the form of ^{4}He. To compute the maximum possible value of Y_p, suppose that every neutron present after the proton-neutron freezeout is incorporated into a ^{4}He nucleus. Given a neutron-to-proton ratio of $n_n/n_p = 1/5$, we can consider a representative group of 2 neutrons and 10 protons. The 2 neutrons can fuse with 2 of the protons to form a single ^{4}He nucleus. The remaining 8 protons, though, will remain unfused. The mass fraction of ^{4}He will then be

$$Y_{\rm max} = \frac{4}{12} = \frac{1}{3}. \tag{10.21}$$

More generally, if $f \equiv n_n/n_p$, with $0 \leq f \leq 1$, then the maximum possible value of Y_p is $Y_{\rm max} = 2f/(1+f)$.

If the observed value of $Y_p = 0.24$ were greater than the predicted $Y_{\rm max}$, that would be a cause for worry; it might mean, for example, that we didn't really understand the process of proton-neutron freezeout. However, the fact that the observed value of Y_p is less than $Y_{\rm max}$ is not worrisome; various factors act to reduce the actual value of Y_p below its theoretical maximum. First, if nucleosynthesis didn't take place immediately after freezeout at $t \approx 1$ s, then the spontaneous decay of neutrons would inevitably lower the neutron-to-proton ratio, and thus reduce the amount of ^{4}He produced. Next, if some neutrons escape fusion altogether, or end in nuclei lighter than ^{4}He (such as D or ^{3}He), they will not contribute to Y_p. Finally, if nucleosynthesis goes on long enough to produce nuclei heavier than ^{4}He, that too will reduce Y_p.

In order to compute Y_p accurately, as well as the abundances of other isotopes, it will be necessary to consider the process of nuclear fusion in more detail. Fortunately, much of the statistical mechanics we will need is just a rehash of what we used when studying recombination.

10.3 ■ DEUTERIUM SYNTHESIS

Let's move on to the next stage of Big Bang Nucleosynthesis, just after proton-neutron freezeout is complete. The time is $t \approx 2$ s. The neutron-to-proton ratio is $n_n/n_p = 0.2$. The neutrinos, which ceased to interact with electrons about the same time they stopped interacting with neutrons and protons, are now decoupled from the rest of the universe. The photons, however, are still strongly coupled to the protons and neutrons. Big Bang Nucleosynthesis takes place through a series of two-body reactions, building heavier nuclei step by step. The essential first step in BBN is the fusion of a proton and a neutron to form a deuterium nucleus:

$$p + n \rightleftharpoons \mathrm{D} + \gamma. \tag{10.22}$$

When a proton and a neutron fuse, the energy released (and carried away by a gamma ray) is the binding energy of a deuterium nucleus:

$$B_{\mathrm{D}} = (m_n + m_p - m_{\mathrm{D}})c^2 = 2.22\,\mathrm{MeV}. \tag{10.23}$$

Conversely, a photon with energy $\geq B_{\mathrm{D}}$ can photodissociate a deuterium nucleus into its component proton and neutron. The reaction shown in equation (10.22) should have a haunting familiarity if you've just read Chapter 9; it has the same structural form as the reaction governing the recombination of hydrogen:

$$p + e^- \rightleftharpoons H + \gamma. \tag{10.24}$$

A comparison of equation (10.22) with equation (10.24) shows that in each case, two particles become bound together to form a composite object, with the excess energy carried away by a photon. In the case of nucleosynthesis, a proton and neutron are bonded together by the strong nuclear force to form a deuterium nucleus, with a gamma-ray photon being emitted. In the case of photoionization, a proton and electron are bonded together by the electromagnetic force to form a neutral hydrogen atom, with an ultraviolet photon being emitted. A major difference between nucleosynthesis and recombination, of course, is between the energy scales involved. The binding energy of deuterium, $B_{\rm D} = 2{,}200{,}000\,{\rm eV}$, is 160,000 times greater than the ionization energy of hydrogen, $Q = 13.6\,{\rm eV}$.[3]

Despite the difference in energy scales, many of the equations used to analyze recombination can be re-used to analyze deuterium nucleosynthesis. Around the time of recombination, for instance, the relative numbers of free protons, free electrons, and neutral hydrogen atoms is given by the Saha equation,

$$\frac{n_H}{n_p n_e} = \left(\frac{m_e kT}{2\pi\hbar^2}\right)^{-3/2} \exp\left(\frac{Q}{kT}\right), \tag{10.25}$$

which tells us that neutral hydrogen is favored in the limit $kT \to 0$, and that ionized hydrogen is favored in the limit $kT \to \infty$. Around the time of deuterium synthesis, the relative numbers of free protons, free neutrons, and deuterium nuclei are given by an equation directly analogous to equation (9.22):

$$\frac{n_{\rm D}}{n_p n_n} = \frac{g_{\rm D}}{g_p g_n}\left(\frac{m_D}{m_p m_n}\right)^{3/2}\left(\frac{kT}{2\pi\hbar^2}\right)^{-3/2} \exp\left(\frac{[m_p + m_n - m_{\rm D}]c^2}{kT}\right). \tag{10.26}$$

From equation (10.23), we can make the substitution $[m_p + m_n - m_{\rm D}]c^2 = B_{\rm D}$. The statistical weight factor of the deuterium nucleus is $g_{\rm D} = 3$, in comparison to $g_p = g_n = 2$ for a proton or neutron. To acceptable accuracy, we may write $m_p = m_n = m_{\rm D}/2$. These substitutions yield the nucleosynthetic equivalent of the Saha equation,

$$\frac{n_{\rm D}}{n_p n_n} = 6\left(\frac{m_n kT}{\pi\hbar^2}\right)^{-3/2} \exp\left(\frac{B_{\rm D}}{kT}\right), \tag{10.27}$$

which tells us that deuterium is favored in the limit $kT \to 0$, and that free protons and neutrons are favored in the limit $kT \to \infty$.

To define a precise temperature $T_{\rm nuc}$ at which the nucleosynthesis of deuterium takes place, we need to define what we mean by "the nucleosynthesis of deuterium." Just as recombination takes a finite length of time, so does nucleosynthesis. It is useful, though, to define $T_{\rm nuc}$ as the temperature at which $n_{\rm D}/n_n = 1$; that is, the temperature at which half the free neutrons have been fused into deuterium nuclei. As long as equation (10.27) holds true, the deuterium-to-neutron ratio can be written as

$$\frac{n_{\rm D}}{n_n} = 6n_p\left(\frac{m_n kT}{\pi\hbar^2}\right)^{-3/2} \exp\left(\frac{B_{\rm D}}{kT}\right). \tag{10.28}$$

We can write the deuterium-to-neutron ratio as a function of T and the baryon-to-photon ratio η if we make some simplifying assumptions. Even today, we know

[3]As the makers of bombs have long known, you can release much more energy by fusing atomic nuclei than by simply shuffling electrons around.

that $\sim$ 75% of all the baryons in the universe are in the form of unbound protons. Before the start of deuterium synthesis, 5 out of 6 baryons (or $\sim$ 83%) were in the form of unbound protons. Thus, if we don't want to be fanatical about accuracy, we can write

$$n_p \approx 0.8 n_{\rm bary} = 0.8 \eta n_\gamma = 0.8\eta \left[0.243 \left(\frac{kT}{\hbar c} \right)^3 \right]. \tag{10.29}$$

Substituting equation (10.29) into equation (10.28), we find that the deuterium-to-neutron ratio is a relatively simple function of temperature:

$$\frac{n_{\rm D}}{n_n} \approx 6.5\eta \left(\frac{kT}{m_n c^2} \right)^{3/2} \exp\left(\frac{B_{\rm D}}{kT} \right). \tag{10.30}$$

This function is plotted in Figure 10.3, assuming a baryon-to-photon ratio of $\eta = 5.5 \times 10^{-10}$. The temperature $T_{\rm nuc}$ of deuterium nucleosynthesis can be found by solving the equation

$$1 \approx 6.5\eta \left(\frac{kT_{\rm nuc}}{m_n c^2} \right)^{3/2} \exp\left(\frac{B_{\rm D}}{kT_{\rm nuc}} \right). \tag{10.31}$$

With $m_n c^2 = 939.6\,\text{MeV}$, $B_{\rm D} = 2.22\,\text{MeV}$, and $\eta = 5.5 \times 10^{-10}$, the temperature of deuterium synthesis is $kT_{\rm nuc} \approx 0.066\,\text{MeV}$, corresponding to $T_{\rm nuc} \approx$

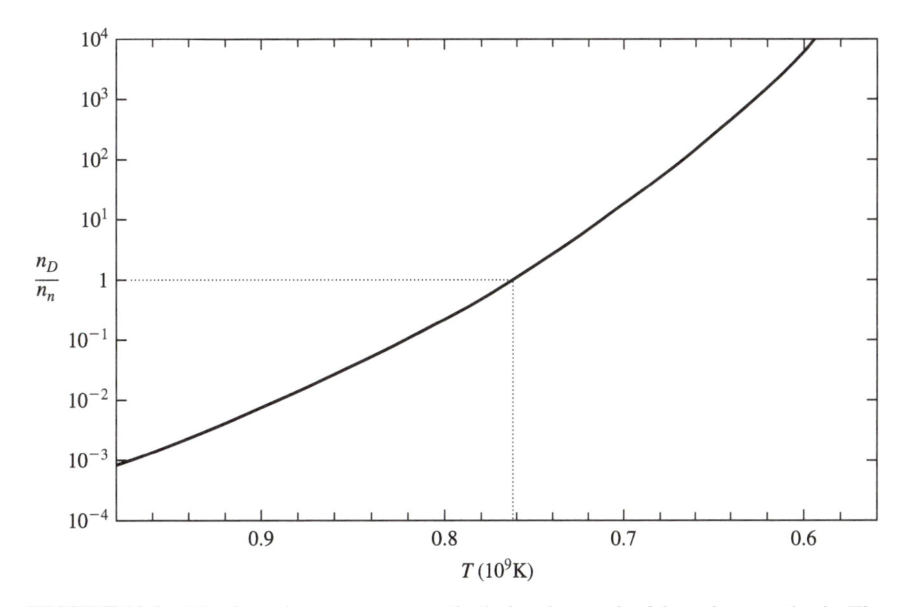

FIGURE 10.3 The deuterium-to-neutron ratio during the epoch of deuterium synthesis. The nucleosynthetic equivalent of the Saha equation (equation (10.27)) is assumed to hold true.

7.6×10^8 K. The temperature drops to this value when the age of the universe is $t_{\rm nuc} \approx 200$ s.

Note that the time delay until the start of nucleosynthesis, $t_{\rm nuc} \approx 200$ s, is not negligible compared to the decay time of the neutron, $\tau_n = 890$ s. By the time nucleosynthesis actually gets underway, neutron decay has slightly decreased the neutron-to-proton ratio from $n_n/n_p = 1/5$ to

$$\frac{n_n}{n_p} \approx \frac{\exp(-200/890)}{5 + [1 - \exp(-200/890)]} \approx \frac{0.8}{5.2} \approx 0.15. \tag{10.32}$$

This in turn lowers the maximum possible ^{4}He mass fraction from $Y_{\rm max} \approx 0.33$ to $Y_{\rm max} \approx 0.27$.

10.4 ■ BEYOND DEUTERIUM

The deuterium-to-neutron ratio $n_{\rm D}/n_n$ does not remain indefinitely at the equilibrium value given by equation (10.30). Once a significant amount of deuterium forms, many possible nuclear reactions are available. For instance, a deuterium nucleus can fuse with a proton to form ^{3}He:

$$\mathrm{D} + p \rightleftharpoons {}^3\mathrm{He} + \gamma. \tag{10.33}$$

Alternatively, it can fuse with a neutron to form ^{3}H, also known as tritium:

$$\mathrm{D} + n \rightleftharpoons {}^3\mathrm{H} + \gamma. \tag{10.34}$$

Tritium is unstable; it spontaneously decays to ^{3}He, emitting an electron and an electron anti-neutrino in the process. However, the decay time of tritium is approximately 18 years; during the brief time that Big Bang Nucleosynthesis lasts, tritium can be regarded as effectively stable.

Deuterium nuclei can also fuse with each other to form ^{4}He:

$$\mathrm{D} + \mathrm{D} \rightleftharpoons {}^4\mathrm{He} + \gamma. \tag{10.35}$$

However, it is more likely that the interaction of two deuterium nuclei will end in the formation of a tritium nucleus (with the emission of a proton),

$$\mathrm{D} + \mathrm{D} \rightleftharpoons {}^3\mathrm{H} + p, \tag{10.36}$$

or the formation of a ^{3}He nucleus (with the emission of a neutron),

$$\mathrm{D} + \mathrm{D} \rightleftharpoons {}^3\mathrm{He} + n. \tag{10.37}$$

A large amount of ^{3}H or ^{3}He is never present during the time of nucleosynthesis. Soon after they are formed, they are converted to ^{4}He by reactions such as

$$
\begin{aligned}
{}^3\mathrm{H} + p &\rightleftharpoons {}^4\mathrm{He} + \gamma \\
{}^3\mathrm{He} + n &\rightleftharpoons {}^4\mathrm{He} + \gamma \\
{}^3\mathrm{H} + \mathrm{D} &\rightleftharpoons {}^4\mathrm{He} + n \\
{}^3\mathrm{He} + \mathrm{D} &\rightleftharpoons {}^4\mathrm{He} + p.
\end{aligned} \tag{10.38}
$$

None of the post-deuterium reactions outlined in equations (10.33) through (10.38) involve neutrinos; they all involve the strong nuclear force, and have large cross-sections and fast reaction rates. Thus, once nucleosynthesis begins, D, ^{3}H, and ^{3}He are all efficiently converted to ^{4}He.

Once ^{4}He is reached, however, the orderly march of nucleosynthesis to heavier and heavier nuclei reaches a roadblock. For such a light nucleus, ^{4}He is exceptionally tightly bound, as illustrated in Figure 10.1. By contrast, there are no stable nuclei with $A = 5$. If you try to fuse a proton or neutron to ^{4}He, it won't work; ^{5}He and ^{5}Li are not stable nuclei. Thus, ^{4}He is resistant to fusion with protons and neutrons. Small amounts of ^{6}Li and ^{7}Li, the two stable isotopes of lithium, are made by reactions such as

$$
{}^4\mathrm{He} + \mathrm{D} \rightleftharpoons {}^6\mathrm{Li} + \gamma \tag{10.39}
$$

and

$$
{}^4\mathrm{He} + {}^3\mathrm{H} \rightleftharpoons {}^7\mathrm{Li} + \gamma. \tag{10.40}
$$

In addition, small amounts of ^{7}Be are made by reactions such as

$$
{}^4\mathrm{He} + {}^3\mathrm{He} \rightleftharpoons {}^7\mathrm{Be} + \gamma. \tag{10.41}
$$

The synthesis of nuclei with $A > 7$ is hindered by the absence of stable nuclei with $A = 8$. For instance, if ^{8}Be is made by the reaction

$$
{}^4\mathrm{He} + {}^4\mathrm{He} \rightarrow {}^8\mathrm{Be}, \tag{10.42}
$$

then the ^{8}Be nucleus falls back apart into a pair of ^{4}He nuclei with a decay time of only $\tau = 3 \times 10^{-16}$ s.

The bottom line is that once deuterium begins to be formed, fusion up to the tightly bound ^{4}He nucleus proceeds very rapidly. Fusion of heavier nuclei occurs much less rapidly. The precise yields of the different isotopes involved in BBN are customarily calculated using a fairly complex computer code. The complexity is necessary because of the large number of possible reactions that can occur once deuterium has been formed, all of which have temperature-dependent cross-sections. Thus, there's a good deal of bookkeeping involved. The results of a typical BBN code, which follows the mass fraction of different isotopes as the universe expands and cools, is shown in Figure 10.4. Initially, at $T \gg 10^9$ K, almost all the baryonic matter is in the form of free protons and neutrons. As the deuterium density climbs upward, however, the point is eventually reached where

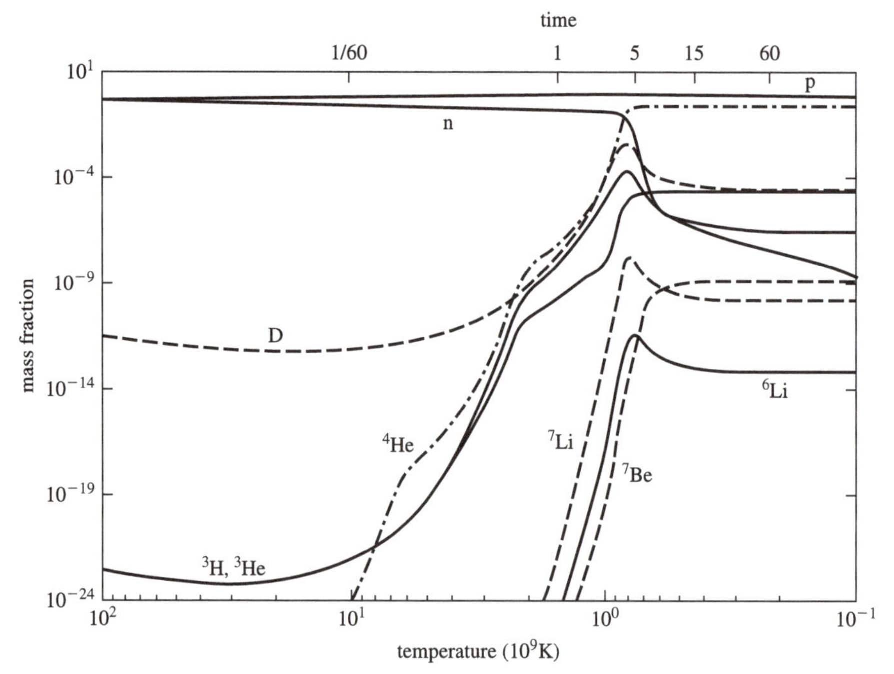

FIGURE 10.4 Mass fraction of nuclei as a function of time during the epoch of nucleosynthesis. A baryon-to-photon ratio of $\eta = 5.1 \times 10^{-10}$ is assumed.

significant amounts of ^{3}H, ^{3}He, and ^{4}He are formed. By the time the temperature has dropped to $T \sim 4 \times 10^8$ K, at $t \sim 10$ min, Big Bang Nucleosynthesis is essentially over. Nearly all the baryons are in the form of free protons or ^{4}He nuclei. The small residue of free neutrons decays into protons. Small amounts of D, ^{3}H, and ^{3}He are left over, a tribute to the incomplete nature of Big Bang Nucleosynthesis. (^{3}H later decays to ^{3}He.) Very small amounts of ^{6}Li, ^{7}Li, and ^{7}Be are made. (^{7}Be is later converted to ^{7}Li by electron capture: $^7\text{Be} + e^- \rightarrow {}^7\text{Li} + \nu_e$.)

The yields of D, ^{3}He, ^{4}He, ^{6}Li, and ^{7}Li depend on various physical parameters. Most importantly, they depend on the baryon-to-photon ratio η. A high baryon-to-photon ratio increases the temperature T_{nuc} at which deuterium synthesis occurs, and hence gives an earlier start to Big Bang Nucleosynthesis. Since BBN is a race against the clock as the density and temperature of the universe drop, getting an earlier start means that nucleosynthesis is more efficient at producing ^{4}He, leaving less D and ^{3}He as leftovers. A plot of the mass fraction of various elements produced by Big Bang Nucleosynthesis is shown in Figure 10.5. Note that larger values of η produce larger values for Y_p (the ^{4}He mass fraction) and smaller values for the deuterium density, as explained above. The dependence of the ^{7}Li density on η is more complicated. Within the range of η plotted in Figure 10.5, the direct production of ^{7}Li by the fusion of ^{4}He and ^{3}H is a decreasing function

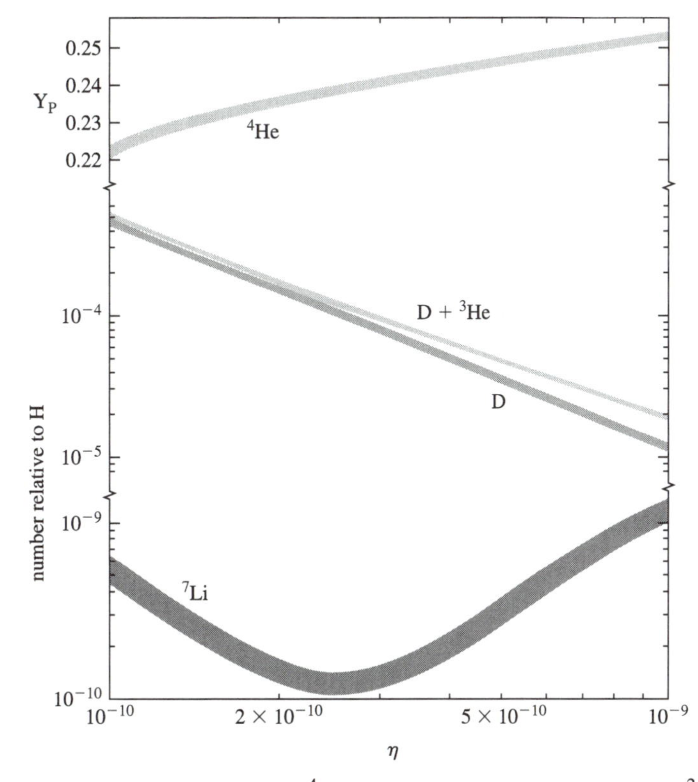

FIGURE 10.5 The mass fraction of ^{4}He, and the number densities of D, D+^{3}He, and ^{7}Li expressed as a fraction of the H number density. The width of each line represents the 95% confidence interval in the density.

of η, while the indirect production of ^{7}Li by ^{7}Be electron capture is an increasing function of η. The net result is a minimum in the predicted density of ^{7}Li at $\eta \approx 3 \times 10^{-10}$.

Broadly speaking, we know immediately that the baryon-to-photon ratio can't be as small as $\eta \sim 10^{-12}$. If it were, BBN would be extremely inefficient, and we would expect only tiny amounts of helium to be produced ($Y_p < 0.01$). Conversely, we know that the baryon-to-photon ratio can't be as large as $\eta \sim 10^{-7}$. If it were, nucleosynthesis would have taken place very early (before neutrons had a chance to decay), the universe would be essentially deuterium-free, and Y_p would be near its maximum permissible value of $Y_{\max} \approx 0.33$. Pinning down the value of η more accurately requires making accurate observations of the *primordial* densities of the light elements; that is, the densities before nucleosynthesis in stars started to alter the chemical composition of the universe. In determining the value of η, it is most useful to determine the primordial abundance of deuterium. This is because the deuterium abundance is strongly dependent on η in the range of interest. Thus, determining the deuterium abundance with only modest accuracy will enable us to determine η fairly well. By contrast, the primordial helium fraction, Y_p, has only a weak dependence on η for the range of interest, as shown in Figure 10.5. Thus, determining η with a fair degree of accuracy would require measuring Y_p with fanatic precision.

A major problem in determining the primordial deuterium abundance is that deuterium is very easily destroyed in stars. When an interstellar gas cloud collapses gravitationally to form a star, the first fusion reactions that occur involve the fusion of deuterium into helium. Thus, the abundance of deuterium in the universe tends to decrease with time.[4] Deuterium abundances are customarily given as the ratio of the number of deuterium atoms to the number of hydrogen atoms (D/H). For instance, in the local interstellar medium within our galaxy, astronomers find an average value $D/H \approx 1.6 \times 10^{-5}$; that is, there's one deuterium atom for every 60,000 ordinary hydrogen atoms. However, the Sun and the interstellar medium are contaminated with material that has been cycled through stellar interiors. Thus, we expect the primordial deuterium-to-hydrogen value to have been $(D/H)_p > 1.6 \times 10^{-5}$.

Currently, the most promising way to find the primordial value of D/H is to look at the spectra of distant quasars. We don't care about the deuterium within the quasar itself; instead, we just want to use the quasar as a flashlight to illuminate the intergalactic gas clouds that lie between it and us. If an intergalactic gas cloud contains no detectable stars, and has very low levels of elements heavier than lithium, we can hope that its D/H value is close to the primordial value, and hasn't been driven downward by the effects of fusion within stars. Neutral hydrogen atoms within these intergalactic clouds will absorb photons whose energy corresponds to the Lyman-α transition; that is, the transition of the atom's electron from the ground state ($n = 1$) to the next higher energy level ($n = 2$). In an ordinary hydrogen atom (^{1}H), the Lyman-α transition corresponds to a wavelength $\lambda_H = 121.57\,\text{nm}$. In a deuterium atom, the greater mass of the nucleus causes a small isotopic shift in the electron's energy levels. As a consequence, the Lyman-α transition in deuterium corresponds to a slightly shorter wavelength, $\lambda_D = 121.54\,\text{nm}$. When we look at light from a quasar that has passed through an intergalactic cloud at redshift z_{cl}, we will see a strong absorption line at $\lambda_H(1 + z_{cl})$, due to absorption from ordinary hydrogen, and a much weaker absorption line at $\lambda_D(1 + z_{cl})$, due to absorption from deuterium. Detailed studies of the strength of the absorption lines in the spectra of different quasars give results consistent with the ratio $(D/H) = (3.0 \pm 0.4) \times 10^{-5}$. This translates into a baryon-to-photon ratio of $\eta = (5.5 \pm 0.5) \times 10^{-10}$.

The value of η can be converted into a value for the current baryon density by the relation

$$n_{\text{bary},0} = \eta n_{\gamma,0} = 0.23 \pm 0.02\,\text{m}^{-3}. \tag{10.43}$$

[4]There are no mechanisms known that will create deuterium in significant amounts after Big Bang Nucleosynthesis is complete.

Since most of the baryons are protons, we can write, to acceptable accuracy,

$$\varepsilon_{\text{bary},0} = (m_p c^2) n_{\text{bary},0} = 210 \pm 20\,\text{MeV}\,\text{m}^{-3}. \tag{10.44}$$

The current density parameter in baryons is thus

$$\Omega_{\text{bary},0} = \frac{\varepsilon_{\text{bary},0}}{\varepsilon_{c,0}} = \frac{210 \pm 20\,\text{MeV}\,\text{m}^{-3}}{5200 \pm 1000\,\text{MeV}\,\text{m}^{-3}} = 0.04 \pm 0.01. \tag{10.45}$$

Note that the largest source of uncertainty in the value of $\Omega_{\text{bary},0}$ is not the uncertainty in the baryon density, but the uncertainty in the critical density (which in turn results from the fact that H_0 is not particularly well determined).

10.5 ■ BARYON–ANTIBARYON ASYMMETRY

The results of Big Bang Nucleosynthesis tell us what the universe was like when it was relatively hot ($T_{\text{nuc}} \approx 7 \times 10^8$ K) and dense:

$$\varepsilon_{\text{nuc}} \approx \alpha T_{\text{nuc}}^4 \approx 10^{33}\,\text{MeV}\,\text{m}^{-3}. \tag{10.46}$$

This energy density corresponds to a mass density of $\varepsilon_{\text{nuc}}/c^2 \approx 2000\,\text{kg}\,\text{m}^{-3}$, or about twice the density of water. Remember, though, that the energy density at the time of BBN was almost entirely in the form of radiation. The mass density of baryons at the time of BBN was

$$\rho_{\text{bary}}(t_{\text{nuc}}) = \Omega_{\text{bary},0}\rho_{c,0}\left(\frac{T_{\text{nuc}}}{T_0}\right)^3 \tag{10.47}$$

$$\approx (0.04)(9.2 \times 10^{-27}\,\text{kg}\,\text{m}^{-3})\left(\frac{7 \times 10^8}{2.725}\right)^3 \approx 0.007\,\text{kg}\,\text{m}^{-3}.$$

A density of several grams per cubic meter is not outlandishly high, by everyday standards; it's equal to the density of the Earth's stratosphere. A mean photon energy of $2.7kT_{\text{nuc}} \approx 0.2\,\text{MeV}$ is not outlandishly high, by everyday standards; you are bombarded with photons of about that energy when you have your teeth x-rayed at the dentist. The physics of Big Bang Nucleosynthesis is well understood.

Some of the initial conditions for Big Bang Nucleosynthesis, however, are rather puzzling. The baryon-to-photon ratio, $\eta \approx 5.5 \times 10^{-10}$, is a remarkably small number; the universe seems to have a strong preference for photons over baryons. It's also worthy of remark that the universe seems to have a strong preference for baryons over antibaryons. The laws of physics demand the presence of antiprotons ($\bar{p}$), containing two "anti-up" quarks and one "anti-down" quark apiece, as well as antineutrons ($\bar{n}$), containing one "anti-up" quark and two "anti-down" quarks apiece.[5] In practice, though, we find that the universe has an ex-

[5]Note that an "anti-up" quark is *not* the same as a "down" quark; nor is "anti-down" equivalent to "up."

tremely large excess of protons and neutrons over antiprotons and antineutrons (and hence an excess of quarks over antiquarks). At the time of Big Bang Nucleosynthesis, the number density of antibaryons ($\bar{n}$ and $\bar{p}$) was tiny compared to the number density of baryons, which in turn was tiny compared to the number density of photons. This imbalance, $n_{\rm antibary} \ll n_{\rm bary} \ll n_\gamma$, has its origin in the physics of the very early universe.

When the temperature of the early universe was greater than $kT \approx 150\,\text{MeV}$, the quarks it contained were not confined within baryons and other particles, as they are today, but formed a sea of free quarks (sometimes referred to by the oddly culinary name of "quark soup"). During the first few microseconds of the universe, when the quark soup was piping hot, quarks and antiquarks were constantly being created by pair production and destroyed by mutual annihilation:

$$\gamma + \gamma \rightleftharpoons q + \bar{q}, \tag{10.48}$$

where q and $\bar{q}$ could represent, for instance, an "up" quark and an "anti-up" quark, or a "down" quark and an "anti-down" quark. During this period of quark pair production, the numbers of "up" quarks, "anti-up" quarks, "down" quarks, "anti-down" quarks, and photons were nearly equal to each other. However, suppose there were a very tiny asymmetry between quarks and antiquarks, such that

$$\delta_q \equiv \frac{n_q - n_{\bar{q}}}{n_q + n_{\bar{q}}} \ll 1. \tag{10.49}$$

As the universe expanded and the quark soup cooled, quark-antiquark pairs would no longer be produced. The existing antiquarks would then annihilate with the quarks. However, because of the small excess of quarks over antiquarks, there would be a residue of quarks with number density

$$\frac{n_q}{n_\gamma} \sim \delta_q. \tag{10.50}$$

Thus, if there were 1,000,000,000 quarks for every 999,999,997 antiquarks in the early universe, three lucky quarks in a billion would be left over after the others encountered antiquarks and were annihilated. The leftover quarks, however, would be surrounded by roughly 2 billion photons, the product of the annihilations. After the three quarks were bound together into a baryon at $kT \approx 150\,\text{MeV}$, the resulting baryon-to-photon ratio would be $\eta \sim 5 \times 10^{-10}$.

Thus, the very strong asymmetry between baryons and antibaryons today and the large number of photons per baryon are both products of a tiny asymmetry between quarks and antiquarks in the early universe. The exact origin of the quark-antiquark asymmetry in the early universe is still not exactly known. The physicist Andrei Sakharov, as far back as 1967, was the first to outline the necessary physical conditions for producing a small asymmetry; however, the precise mechanism by which the quarks first developed their few-parts-per-billion advantage over antiquarks still remains to be found.

SUGGESTED READING

Full references are given in the Annotated Bibliography on page 235.

Bernstein (1995), ch. 4, 5, 6: Puts nucleosynthesis in the larger context of particle physics and thermodynamics

Weinberg (1993): Gives some of the historical background to the development of Big Bang Nucleosynthesis

PROBLEMS

10.1. Suppose the neutron decay time were $\tau_n = 89\,\text{s}$ instead of $\tau_n = 890\,\text{s}$, with all other physical parameters unchanged. Estimate Y_{max}, the maximum possible mass fraction in ^{4}He, assuming that all available neutrons are incorporated into ^{4}He nuclei.

10.2. Suppose the difference in rest energy of the neutron and proton were $Q_n = (m_n - m_p)c^2 = 0.129\,\text{MeV}$ instead of $Q_n = 1.29\,\text{MeV}$, with all other physical parameters unchanged. Estimate Y_{max}, the maximum possible mass fraction in ^{4}He, assuming that all available neutrons are incorporated into ^{4}He nuclei.

10.3. A fascinating bit of cosmological history is that of George Gamow's prediction of the Cosmic Microwave Background in 1948. (Unfortunately, his prediction was premature; by the time the CMB was actually discovered in the 1960's, his prediction had fallen into obscurity.) Let's see if you can reproduce Gamow's line of argument. Gamow knew that nucleosynthesis must have taken place at a temperature $T_{\text{nuc}} \approx 10^9\,\text{K}$, and that the age of the universe is currently $t_0 \approx 10\,\text{Gyr}$.

Assume that the universe is flat and contains only radiation. With these assumptions, what was the energy density ε at the time of nucleosynthesis? What was the Hubble parameter H at the time of nucleosynthesis? What was the time t_{nuc} at which nucleosynthesis took place? What is the current temperature T_0 of the radiation filling the universe today? If the universe switched from being radiation-dominated to being matter-dominated at a redshift $z_{rm} > 0$, will this increase or decrease T_0 for fixed values of T_{nuc} and t_0? Explain your answer.

10.4. The total luminosity of the stars in our galaxy is $L \approx 2.3 \times 10^{10}\,\text{L}_\odot$. Suppose that the luminosity of our galaxy has been constant for the past 10 Gyr. How much energy has our galaxy emitted in the form of starlight during that time? Most stars are powered by the fusion of H into ^{4}He, with the release of 28.4 MeV for every helium nucleus formed. How many helium nuclei have been created within stars in our galaxy over the course of the past 10 Gyr, assuming that the fusion of H into ^{4}He is the only significant energy source? If the baryonic mass of our galaxy is $M \approx 10^{11}\,\text{M}_\odot$, by what amount has the helium fraction Y of our galaxy been increased over its primordial value $Y_4 = 0.24$?

10.5. In section 10.2, it is asserted that the maximum possible value of the primordial helium fraction is

$$Y_{\text{max}} = \frac{2f}{1+f}, \tag{10.51}$$

where $f = n_n/n_p \leq 1$ is the neutron-to-proton ratio at the time of nucleosynthesis. Prove that this assertion is true.

10.6. The typical energy of a neutrino in the Cosmic Neutrino Background, as pointed out in Chapter 5, is $E_\nu \sim kT_\nu \sim 5 \times 10^{-4}\,\text{eV}$. What is the approximate interaction cross-section σ_w for one of these cosmic neutrinos? Suppose you had a large lump of ^{56}Fe (with density $\rho = 7900\,\text{kg}\,\text{m}^{-3}$). What is the number density of protons, neutrons, and electrons within the lump of iron? How far, on average, would a cosmic neutrino travel through the iron before interacting with a proton, neutron, or electron? (Assume that the cross-section for interaction is simply σ_w, regardless of the type of particle the neutrino interacts with.)

CHAPTER 11

Inflation and the Very Early Universe

The observed properties of galaxies, quasars, and supernovae at relatively small redshift ($z < 6$) tell us about the universe at times $t > 1\,\text{Gyr}$. The properties of the Cosmic Microwave Background tell us about the universe at the time of last scattering ($z_{\text{ls}} \approx 1100$, $t_{\text{ls}} \approx 0.35\,\text{Myr}$). The abundances of light elements such as deuterium and helium tell us about the universe at the time of Big Bang Nucleosynthesis ($z_{\text{nuc}} \approx 3 \times 10^8$, $t_{\text{nuc}} \approx 3\,\text{min}$). In fact, the observation that primordial gas clouds are roughly one-fourth helium by mass, rather than being all helium or all hydrogen, tells us that we have a fair understanding of what was happening at the time of neutron-proton freezeout ($z_{\text{freeze}} \approx 4 \times 10^9$, $t_{\text{freeze}} \approx 1\,\text{s}$).

So far, our description of the Hot Big Bang scenario has been a triumphal progress, with only minor details (such as the exact determination of H_0, $\Omega_{m,0}$, and $\Omega_{\Lambda,0}$) remaining to be worked out. Whenever a conquering general made a triumphal progress into ancient Rome, a slave stood behind him, whispering in his ear, "Remember, you are mortal." Just as every triumphant general must be reminded of his flawed, imperfect nature, lest he become insufferably arrogant, so every triumphant theory should be inspected carefully for flaws and imperfections. The standard Hot Big Bang scenario, in which the early universe was radiation-dominated, has three underlying problems, called the *flatness problem*, the *horizon problem*, and the *monopole problem*. The flatness problem can be summarized by the statement, "The universe is nearly flat today, and was even flatter in the past." The horizon problem can be summarized by the statement, "The universe is nearly isotropic and homogeneous today, and was even more so in the past." The monopole problem can be summarized by the statement, "The universe is apparently free of magnetic monopoles." To see why these simple statements pose a problem to the standard Hot Big Bang scenario, it is necessary to go a little deeper into the physics of the expanding universe.

11.1 ■ THE FLATNESS PROBLEM

Let's start by examining the flatness problem. The spatial curvature of the universe is related to the density parameter Ω by the Friedmann equation:

$$1 - \Omega(t) = -\frac{\kappa c^2}{R_0^2 a(t)^2 H(t)^2}. \tag{11.1}$$

(Here we use the Friedmann equation in the form given by equation (4.29).) At the present moment, the density parameter and curvature are linked by the equation

$$1 - \Omega_0 = -\frac{\kappa c^2}{R_0^2 H_0^2}. \tag{11.2}$$

The results of the type Ia supernova observations and the measurements of the CMB anisotropy are consistent with the value (see Figure 9.7)

$$|1 - \Omega_0| \leq 0.2. \tag{11.3}$$

Why should the value of Ω_0 be so close to one today? It could have had, for instance, the value $\Omega_0 = 10^{-6}$ or $\Omega_0 = 10^6$ without violating any laws of physics. We could of course invoke coincidence by saying that the initial conditions for the universe just happened to produce $\Omega_0 \approx 1$ today. After all, $\Omega_0 = 0.8$ or $\Omega_0 = 1.2$ aren't *that* close to one. However, when you extrapolate the value of $\Omega(t)$ backward into the past, the closeness of Ω to unity becomes more and more difficult to dismiss as a coincidence.

By combining equations (11.1) and (11.2), we find the equation that gives the density parameter as a function of time:

$$1 - \Omega(t) = \frac{H_0^2(1 - \Omega_0)}{H(t)^2 a(t)^2}. \tag{11.4}$$

When the universe was dominated by radiation and matter, at times $t \ll t_{m\Lambda} \approx 10\,\text{Gyr}$, the Hubble parameter was given by equation (6.35):

$$\frac{H(t)^2}{H_0^2} = \frac{\Omega_{r,0}}{a^4} + \frac{\Omega_{m,0}}{a^3}. \tag{11.5}$$

Thus, the density parameter evolved at the rate

$$1 - \Omega(t) = \frac{(1 - \Omega_0)a^2}{\Omega_{r,0} + a\Omega_{m,0}}. \tag{11.6}$$

During the period when the universe was dominated by radiation and matter, the deviation of Ω from one was constantly growing. During the radiation-dominated phase,

$$|1 - \Omega|_r \propto a^2 \propto t. \tag{11.7}$$

During the later matter-dominated phase,

$$|1 - \Omega|_m \propto a \propto t^{2/3}. \tag{11.8}$$

Suppose, as the available evidence indicates, that the universe is described by a model close to the Benchmark Model, with $\Omega_{r,0} = 8.4 \times 10^{-5}$, $\Omega_{m,0} = 0.3 \pm 0.1$, and $\Omega_{\Lambda,0} = 0.7 \pm 0.1$. At the present, therefore, the density parameter falls within the limits $|1 - \Omega_0| \leq 0.2$. At the time of radiation-matter equality, the density parameter Ω_{rm} was equal to one with an accuracy

$$|1 - \Omega_{rm}| \leq 2 \times 10^{-4}. \tag{11.9}$$

If we extrapolate backward to the time of Big Bang Nucleosynthesis, at $a_{\rm nuc} \approx 3.6 \times 10^{-8}$, the deviation of the density parameter $\Omega_{\rm nuc}$ from one was only

$$|1 - \Omega_{\rm nuc}| \leq 3 \times 10^{-14}. \tag{11.10}$$

At the time that deuterium was forming, the density of the universe was equal to the critical density to an accuracy of one part in 30 trillion. If we push our extrapolation as far back as we dare, to the Planck time at $t_P \approx 5 \times 10^{-44}$ s, $a_P \approx 2 \times 10^{-32}$, we find that the density parameter Ω_P was extraordinarily close to one:

$$|1 - \Omega_P| \leq 1 \times 10^{-60}. \tag{11.11}$$

The number 10^{-60} is, of course, very tiny. To use an analogy, in order to change the Sun's mass by one part in 10^{60}, you would have to add or subtract two electrons. Our very existence depends on the fanatically close balance between the actual density and the critical density in the early universe. If, for instance, the deviation of Ω from one at the time of nucleosynthesis had been one part in 30 thousand instead of one part in 30 trillion, the universe would have collapsed in a Big Crunch or expanded to a low-density Big Chill after only a few years. In that case, galaxies, stars, planets, and cosmologists would not have had time to form.

You might try to dismiss the extreme flatness of the the early universe as a coincidence, by saying, "Ω_P might have had any value, but it just happened to be 1 ± 10^{-60}." However, a coincidence at the level of one part in 10^{60} is *extremely* far-fetched. It would be far more satisfactory if we could find a physical mechanism for flattening the universe early in its history, instead of relying on extremely contrived initial conditions at the Planck time.

11.2 ■ THE HORIZON PROBLEM

The "flatness problem," the remarkable closeness of Ω to one in the early universe, is puzzling. It is accompanied, however, by the "horizon problem," which is, if anything, even more puzzling. The horizon problem is simply the statement that the universe is nearly homogeneous and isotropic on very large scales. Why should we regard this as a problem? So far, we've treated the homogeneity and isotropy of the universe as a blessing rather than a curse. It's the homogeneity and isotropy of the universe, after all, that permit us to describe its curvature by the relatively simple Robertson–Walker metric, and its expansion by the relatively simple Friedmann equation. If the universe were inhomogeneous and anisotropic on large scales, it would be much more difficult to describe mathematically.

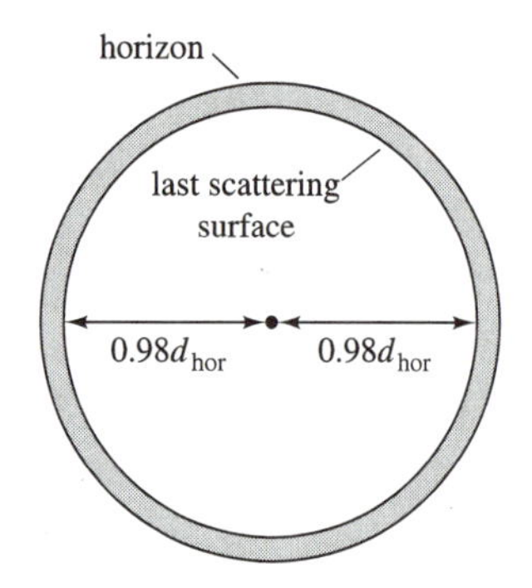

FIGURE 11.1 In the standard Hot Big Bang scenario, the current proper distance to the last scattering surface is 98% of the current horizon distance.

The universe, however, is under no obligation to make things simple for cosmologists. To see why the large scale homogeneity and isotropy of the universe is so unexpected in the standard Hot Big Bang scenario, consider two antipodal points on the last scattering surface, as illustrated in Figure 11.1. The current proper distance to the last scattering surface is

$$d_p(t_0) = c\int_{t_{\rm ls}}^{t_0} \frac{dt}{a(t)}. \tag{11.12}$$

Since the last scattering of the CMB photons occurred a long time ago ($t_{\rm ls} \ll t_0$), the current proper distance to the last scattering surface is only slightly smaller than the current horizon distance. For the Benchmark Model, the current proper distance to the last scattering surface is $d_p(t_0) = 0.98 d_{\rm hor}(t_0)$. Thus, two antipodal points on the last scattering surface, separated by 180° as seen by an observer on Earth, are currently separated by a proper distance of $1.96 d_{\rm hor}(t_0)$. Since the two points are farther apart than the horizon distance, they are causally disconnected. That is, they haven't had time to send messages to each other, and in particular, haven't had time to come into thermal equilibrium with each other. Nevertheless, the two points have the same temperature (once the dipole distortion is subtracted) to within one part in 10^5. Why? How can two points that haven't had time to swap information be so nearly identical in their properties?

The near-isotropy of the Cosmic Microwave Background is still more remarkable when we recall that the temperature fluctuations in the CMB result from the density and velocity fluctuations that existed at the time of last scattering. In the standard Hot Big Bang scenario, the universe was matter-dominated at the time of last scattering, so the horizon distance at that time can be approximated by the value

$$d_{\rm hor}(t_{\rm ls}) = 2\frac{c}{H(t_{\rm ls})} \tag{11.13}$$

appropriate to a flat, matter-only universe (see section 5.4). Since the Hubble distance at the time of last scattering was $c/H(t_{\rm ls}) \approx 0.2\,{\rm Mpc}$, the horizon distance at that time was only $d_{\rm hor}(t_{\rm ls}) \approx 0.4\,{\rm Mpc}$. Thus, points more than 0.4 megaparsecs apart at the time of last scattering were not in causal contact in the standard Hot Big Bang scenario. The angular-diameter distance to the last scattering surface is $d_A \approx 13\,{\rm Mpc}$, as computed in section 9.4. Thus, points on the last scattering surface that were separated by a horizon distance will have an angular separation equal to

$$\theta_{\rm hor} = \frac{d_{\rm hor}(t_{\rm ls})}{d_A} \approx \frac{0.4\,{\rm Mpc}}{13\,{\rm Mpc}} \approx 0.03\,{\rm rad} \approx 2^\circ \tag{11.14}$$

as seen from the Earth today. Therefore, points on the last scattering surface separated by an angle as small as $\sim 2^\circ$ were out of contact with each other at the time the temperature fluctuations were stamped upon the CMB. Nevertheless, we find that $\delta T/T$ is as small as 10^{-5} on scales $\theta > 2^\circ$ (corresponding to $l < 100$ in Figure 9.6).

Why should regions that were out of causal contact with each other at $t_{\rm ls}$ have been so nearly homogeneous in their properties? Invoking coincidence ("The different patches just happened to have the same temperature") requires a great stretch of the imagination. The surface of last scattering can be divided into some 20,000 patches, each two degrees across. In the standard Hot Big Bang scenario, the center of each of these patches was out of touch with the other patches at the time of last scattering. Now, if you invite two people to a potluck dinner, and they both bring potato salad, you can dismiss that as coincidence, even if they had 10^5 different dishes to choose from. However, if you invite 20,000 people to a potluck dinner, and they all bring potato salad, it starts to dawn on you that they must have been in contact with each other: "Psst ... let's all bring potato salad. Pass it on." Similarly, it starts to dawn on you that the different patches of the last scattering surface, in order to be so nearly equal in temperature, must have been in contact with each other: "Psst ... let's all be at $T = 2.725\,{\rm K}$ when the universe is 13.5 gigayears old. Pass it on."

11.3 ■ THE MONOPOLE PROBLEM

The monopole problem—that is, the apparent lack of magnetic monopoles in the universe—is not a purely cosmological problem, but one that results from combining the Hot Big Bang scenario with the particle physics concept of a Grand Unified Theory. In particle physics, a Grand Unified Theory, or GUT, is a field theory that attempts to unify the electromagnetic force, the weak nuclear force, and the strong nuclear force. Unification of forces has been a goal of scientists since the 1870s, when James Clerk Maxwell demonstrated that electricity and magnetism are both manifestations of a single underlying electromagnetic field. Currently, it is customary to speak of the four fundamental forces of nature: gravitational, electromagnetic, weak, and strong. In the view of many physicists, though, four forces are three too many; they've spent much time and effort to show that two or more of the "fundamental forces" are actually different aspects of a single underlying force. About a century after Maxwell, Steven Weinberg, Abdus Salam, and Sheldon Glashow successfully devised an electroweak theory. They demonstrated that at particle energies greater than $E_{\rm ew} \sim 1\,{\rm TeV}$, the electromagnetic force and the weak force unite to form a single "electroweak" force. The electroweak energy of $E_{\rm ew} \sim 1\,{\rm TeV}$ corresponds to a temperature $T_{\rm ew} \sim E_{\rm ew}/k \sim 10^{16}\,{\rm K}$; the universe had this temperature when its age was $t_{\rm ew} \sim 10^{-12}\,{\rm s}$. Thus, when the universe was less than a picosecond old, there were only three fundamental forces: the gravitational, strong, and electroweak force. When the predictions of the electroweak energy were confirmed experimentally, Weinberg, Salam, and Glashow toted home their Nobel Prizes, and physicists braced themselves for the next step: unifying the electroweak force with the strong force.

By extrapolating the known properties of the strong and electroweak forces to higher particle energies, physicists estimate that at an energy $E_{\rm GUT}$ of roughly $10^{12} \rightarrow 10^{13}\,{\rm TeV}$, the strong and electroweak forces should be unified as a single Grand Unified Force. If the GUT energy is $E_{\rm GUT} \sim 10^{12}\,{\rm TeV}$, this corresponds to a temperature $T_{\rm GUT} \sim 10^{28}\,{\rm K}$ and an age for the universe of $t_{\rm GUT} \sim 10^{-36}\,{\rm s}$. The GUT energy is about four orders of magnitude smaller than the Planck energy, $E_P \sim 10^{16}\,{\rm TeV}$. Physicists are searching for a Theory of Everything (TOE) that describes how the Grand Unified Force and the force of gravity ultimately unite to form a single unified force at the Planck scale. The different unification energy scales, and the corresponding temperatures and times in the early universe, are shown in Figure 11.2.

One of the predictions of Grand Unified Theories is that the universe underwent a *phase transition* as the temperature dropped below the GUT temperature. Generally speaking, phase transitions are associated with a spontaneous loss of symmetry as the temperature of a system is lowered. Take, as an example, the phase transition known as "freezing water." At temperatures $T > 273\,{\rm K}$, water is liquid. Individual water molecules are randomly oriented, and the liquid water thus has rotational symmetry about any point; in other words, it is isotropic. However, when the temperature drops below $T = 273\,{\rm K}$, the water undergoes a phase transition, from liquid to solid, and the rotational symmetry of the water

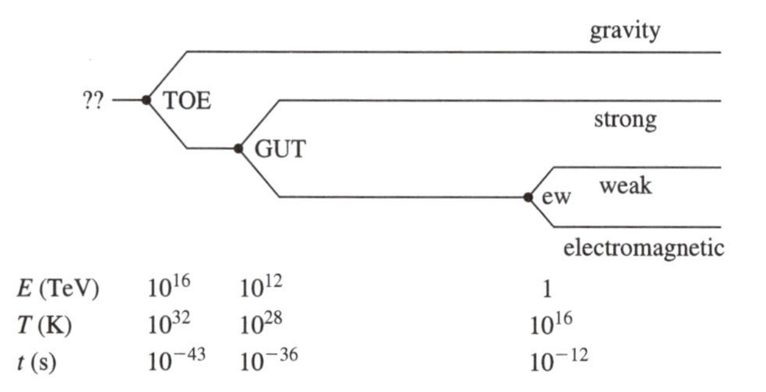

FIGURE 11.2 The energy, temperature, and time scales at which the different force unifications occur.

is lost. The water molecules are locked into a crystalline structure, and the ice no longer has rotational symmetry about an arbitrary point. In other words, the ice crystal is anisotropic, with preferred directions corresponding to the crystal's axes of symmetry.[1] In a broadly similar vein, there is a loss of symmetry when the universe undergoes the GUT phase transition at $t_{\rm GUT} \sim 10^{-36}\,{\rm s}$. At $T > T_{\rm GUT}$, there was a symmetry between the strong and electroweak forces. At $T < T_{\rm GUT}$, the symmetry is spontaneously lost; the strong and electroweak forces begin to behave quite differently from each other.

In general, phase transitions associated with a loss of symmetry give rise to flaws known as *topological defects*. To see how topological defects form, consider a large tub of water cooled below $T = 273\,{\rm K}$. Usually, the freezing of the water will start at two or more widely separated nucleation sites. The crystal that forms about any given nucleation site is very regular, with well-defined axes of symmetry. However, the axes of symmetry of two adjacent ice crystals will be misaligned. At the boundary of two adjacent crystals, there will be a two-dimensional topological defect, called a *domain wall*, where the axes of symmetry fail to line up. Other types of phase transitions give rise to one-dimensional, or line-like, topological defects (in a cosmological context, these linear defects are known as *cosmic strings*). Still other types of phase transitions give rise to zero-dimensional, or point-like, topological defects. Grand Unified Theories predict that the GUT phase transition creates point-like topological defects that act as *magnetic monopoles*. That is, they act as the isolated north pole or south pole of a magnet. The rest energy of the magnetic monopoles created in the GUT phase transition is predicted to be $m_{\rm M}c^2 \sim E_{\rm GUT} \sim 10^{12}\,{\rm TeV}$. This corresponds to a mass of over a nanogram (comparable to that of a bacterium), which is a lot of mass for a single particle to be carrying around. At the time of the GUT phase transition, points further apart than the horizon size will be out of causal contact with each other. Thus, we expect roughly one topological defect per horizon volume, due to the mismatch of fields that are not causally linked. The number density of magnetic monopoles, at the time of their creation, would be

$$n_{\rm M}(t_{\rm GUT}) \sim \frac{1}{(2ct_{\rm GUT})^3} \sim 10^{82}\,{\rm m}^{-3}, \qquad (11.15)$$

and their energy density would be

$$\varepsilon_{\rm M}(t_{\rm GUT}) \sim (m_{\rm M}c^2)n_{\rm M} \sim 10^{94}\,{\rm TeV\,m}^{-3}. \qquad (11.16)$$

This is a large energy density, but it is smaller by ten orders of magnitude than the energy density of radiation at the time of the GUT phase transition:

$$\varepsilon_\gamma(t_{\rm GUT}) \approx \alpha T_{\rm GUT}^4 \sim 10^{104}\,{\rm TeV\,m}^{-3}. \qquad (11.17)$$

[1] Suppose, for instance, that the water freezes in the familiar six-pointed form of a snowflake. It is now only symmetric with respect to rotations of 60° (or integral multiples of that angle) about the snowflake's center.

Thus, the magnetic monopoles wouldn't have kept the universe from being radiation-dominated at the time of the GUT phase transition. However, the magnetic monopoles, being so massive, would soon have become highly non-relativistic, with energy density $\varepsilon_M \propto a^{-3}$. The energy density in radiation, though, was falling off at the rate $\varepsilon_\gamma \propto a^{-4}$. Thus, the magnetic monopoles would have dominated the energy density of the universe when the scale factor had grown by a factor $\sim 10^{10}$; that is, when the temperature had fallen to $T \sim 10^{-10} T_{\rm GUT} \sim 10^{18}$ K, and the age of the universe was only $t \sim 10^{-16}$ s.

Obviously, the universe is *not* dominated by magnetic monopoles today. In fact, there is no strong evidence that they exist at all. Every north magnetic pole we can find is paired with a south magnetic pole, and vice versa. There are no isolated north poles or isolated south poles. The monopole problem can thus be rephrased as the question, "Where have all the magnetic monopoles gone?" Now, you can always answer the question by saying, "There were never any monopoles to begin with." There is not yet a single, definitive Grand Unified Theory, and in some variants on the GUT theme, magnetic monopoles are not produced. However, the flatness and horizon problems are not so readily dismissed. When the physicist Alan Guth first proposed the idea of *inflation* in 1981, he introduced it as a way of resolving the flatness problem, the horizon problem, and the monopole problem with a single cosmological mechanism.

11.4 ■ THE INFLATION SOLUTION

What is inflation? In a cosmological context, inflation can most generally be defined as the hypothesis that there was a period, early in the history of our universe, when the expansion was accelerating outward; that is, an epoch when $\ddot{a} > 0$. The acceleration equation,

$$\frac{\ddot{a}}{a} = -\frac{4\pi G}{3c^2}(\varepsilon + 3P), \tag{11.18}$$

tells us that $\ddot{a} > 0$ when $P < -\frac{\varepsilon}{3}$. Thus, inflation would have taken place if the universe were temporarily dominated by a component with equation-of-state parameter $w < -\frac{1}{3}$. The usual implementation of inflation states that the universe was temporarily dominated by a positive cosmological constant Λ_i (with $w = -1$), and thus had an acceleration equation that could be written in the form

$$\frac{\ddot{a}}{a} = \frac{\Lambda_i}{3} > 0. \tag{11.19}$$

In an inflationary phase when the energy density was dominated by a cosmological constant, the Friedmann equation was

$$\left(\frac{\dot{a}}{a}\right)^2 = \frac{\Lambda_i}{3}. \tag{11.20}$$

The Hubble constant H_i during the inflationary phase was thus constant, with the value $H_i = (\Lambda_i/3)^{1/2}$, and the scale factor grew exponentially with time:

$$a(t) \propto e^{H_i t}. \tag{11.21}$$

To see how a period of exponential growth can resolve the flatness, horizon, and monopole problems, suppose that the universe had a period of exponential expansion sometime in the midst of its early, radiation-dominated phase. For simplicity, suppose the exponential growth was switched on instantaneously at a time t_i, and lasted until some later time t_f, when the exponential growth was switched off instantaneously, and the universe reverted to its former state of radiation-dominated expansion. In this simple case, we can write the scale factor as

$$a(t) = \begin{cases} a_i(t/t_i)^{1/2} & t < t_i \\ a_i e^{H_i(t-t_i)} & t_i < t < t_f \\ a_i e^{H_i(t_f-t_i)}(t/t_f)^{1/2} & t > t_f. \end{cases} \tag{11.22}$$

Thus, between the time t_i, when the exponential inflation began, and the time t_f, when the inflation stopped, the scale factor increased by a factor

$$\frac{a(t_f)}{a(t_i)} = e^N, \tag{11.23}$$

where N, the number of e-foldings of inflation, was

$$N \equiv H_i(t_f - t_i). \tag{11.24}$$

If the duration of inflation, $t_f - t_i$, was long compared to the Hubble time during inflation, H_i^{-1}, then N was large, and the growth in scale factor during inflation was enormous.

For concreteness, let's take one possible model for inflation. This model states that exponential inflation started around the GUT time, $t_i \approx t_{\rm GUT} \approx 10^{-36}$ s, with a Hubble parameter $H_i \approx t_{\rm GUT}^{-1} \approx 10^{36}\,{\rm s}^{-1}$, and lasted for $N \sim 100$ Hubble times. In this particular model, the growth in scale factor during inflation was

$$\frac{a(t_f)}{a(t_i)} \sim e^{100} \sim 10^{43}. \tag{11.25}$$

Note that the cosmological constant Λ_i present at the time of inflation was very large compared to the cosmological constant that seems to be present today. Currently, the evidence is consistent with an energy density in Λ of $\varepsilon_{\Lambda,0} \approx 0.7\varepsilon_{c,0} \approx 0.004\,{\rm TeV\,m^{-3}}$. To produce exponential expansion with a Hubble parameter $H_i \approx 10^{36}\,{\rm s}^{-1}$, the cosmological constant during inflation would have had an energy density (see equation (4.65))

$$\varepsilon_{\Lambda_i} = \frac{c^2}{8\pi G}\Lambda_i = \frac{3c^2}{8\pi G}H_i^2 \sim 10^{105}\,{\rm TeV\,m^{-3}}, \tag{11.26}$$

over 107 orders of magnitude larger.

How does inflation resolve the flatness problem? Equation (11.1), which gives Ω as a function of time, can be written in the form

$$|1 - \Omega(t)| = \frac{c^2}{R_0^2 a(t)^2 H(t)^2} \tag{11.27}$$

for any universe not perfectly flat. If the universe is dominated by a single component with equation-of-state parameter $w \neq -1$, then $a \propto t^{2/(3+3w)}$, $H \propto t^{-1}$, and

$$|1 - \Omega(t)| \propto t^{2(1+3w)/(3+3w)}. \tag{11.28}$$

Thus, if $w < -\frac{1}{3}$, the difference between Ω and one decreases with time. If the universe is expanding exponentially during the inflationary epoch, then

$$|1 - \Omega(t)| \propto e^{-2H_i t}, \tag{11.29}$$

and the difference between Ω and one decreases exponentially with time. If we compare the density parameter at the beginning of exponential inflation ($t = t_i$) with the density parameter at the end of inflation ($t = t_f = t_i + N/H_i$), we find

$$|1 - \Omega(t_f)| = e^{-2N}|1 - \Omega(t_i)|. \tag{11.30}$$

Suppose that prior to inflation, the universe was actually fairly strongly curved, with

$$|1 - \Omega(t_i)| \sim 1. \tag{11.31}$$

After a hundred e-foldings of inflation, the deviation of Ω from one would be

$$|1 - \Omega(t_f)| \sim e^{-2N} \sim e^{-200} \sim 10^{-87}. \tag{11.32}$$

Even if the universe at t_i wasn't particularly close to being flat, a hundred e-foldings of inflation would flatten it like the proverbial pancake. The current limits on the density parameter, $|1 - \Omega_0| \leq 0.2$, imply that $N > 60$, if inflation took place around the GUT time. However, it's possible that N may have been very much greater than 60, since the observational data are entirely consistent with $|1 - \Omega_0| \ll 1$.

How does inflation resolve the horizon problem? At any time t, the horizon distance $d_{\rm hor}(t)$ is given by the relation

$$d_{\rm hor}(t) = a(t)c\int_0^t \frac{dt}{a(t)}. \tag{11.33}$$

Prior to the inflationary period, the universe was radiation-dominated. Thus, the horizon time at the beginning of inflation was

$$d_{\rm hor}(t_i) = a_i c \int_0^{t_i} \frac{dt}{a_i(t/t_i)^{1/2}} = 2ct_i. \tag{11.34}$$

The horizon size at the end of inflation was

$$d_{\rm hor}(t_f) = a_i e^N c \left(\int_0^{t_i} \frac{dt}{a_i (t/t_i)^{1/2}} + \int_{t_i}^{t_f} \frac{dt}{a_i \exp[H_i(t - t_i)]} \right). \tag{11.35}$$

If N, the number of e-foldings of inflation, is large, then the horizon size at the end of inflation was

$$d_{\rm hor}(t_f) = e^N c(2t_i + H_i^{-1}). \tag{11.36}$$

An epoch of exponential inflation causes the horizon size to grow exponentially. If inflation started at $t_i \approx 10^{-36}$ s, with a Hubble parameter $H_i \approx 10^{36}\,{\rm s}^{-1}$, and lasted for $N \approx 100$ e-foldings, then the horizon size immediately before inflation was

$$d_{\rm hor}(t_i) = 2ct_i \approx 6 \times 10^{-28}\,{\rm m}. \tag{11.37}$$

The horizon size immediately after inflation was

$$d_{\rm hor}(t_f) \approx e^N 3ct_i \approx 2 \times 10^{16}\,{\rm m} \approx 0.8\,{\rm pc}. \tag{11.38}$$

During the brief period of $\sim 10^{-34}$ s that inflation lasts in this model, the horizon size is boosted exponentially from submicroscopic scales to nearly a parsec. At the end of the inflationary epoch, the horizon size reverts to growing at a sedate linear rate.

The net result of inflation is to increase the horizon length in the post-inflationary universe by a factor $\sim e^N$ over what it would have been without inflation. For instance, we found that, in the absence of inflation, the horizon size at the time of last scattering was $d_{\rm hor}(t_{\rm ls}) \approx 0.4$ Mpc. Given a hundred e-foldings of inflation in the early universe, however, the horizon size at last scattering would have been $\sim 10^{43}$ Mpc, obviously gargantuan enough for the entire last scattering surface to be in causal contact.

To look at the resolution of the horizon problem from a slightly different viewpoint, consider the entire universe directly visible to us today, that is, the region bounded by the surface of last scattering. Currently, the proper distance to the surface of last scattering is

$$d_p(t_0) \approx 1.4 \times 10^4\,{\rm Mpc}. \tag{11.39}$$

If inflation ended at $t_f \sim 10^{-34}$ s, that corresponds to a scale factor $a_f \sim 2 \times 10^{-27}$. Thus, immediately after inflation, the portion of the universe currently visible to us was crammed into a sphere of proper radius

$$d_p(t_f) = a_f d_p(t_0) \sim 3 \times 10^{-23}\,{\rm Mpc} \sim 0.9\,{\rm m}. \tag{11.40}$$

Immediately after inflation, in this model, all the mass-energy destined to become the hundreds of billions of galaxies we see today was enclosed within a sphere only six feet across. But, to quote Al Jolson, you ain't heard nothin' yet, folks.

If there were $N = 100$ e-foldings of inflation, then immediately prior to the inflationary epoch, the currently visible universe was enclosed within a sphere of proper radius

$$d_p(t_i) = e^{-N} d_p(t_f) \sim 3 \times 10^{-44}\,\mathrm{m}. \tag{11.41}$$

Note that this distance is 16 orders of magnitude smaller than the horizon size immediately prior to inflation ($d_{\mathrm{hor}}(t_i) \sim 6 \times 10^{-28}\,\mathrm{m}$). Thus, the portion of the universe we can see today had plenty of time to achieve thermal uniformity before inflation began.

How does inflation resolve the monopole problem? If magnetic monopoles were created before or during inflation, then the number density of monopoles was diluted to an undetectably low level. During a period when the universe was expanding exponentially ($a \propto e^{H_i t}$), the number density of monopoles, if they were neither created nor destroyed, was decreasing exponentially ($n_{\mathrm{M}} \propto e^{-3H_i t}$). For instance, if inflation started around the GUT time, when the number density of magnetic monopoles was $n_{\mathrm{M}}(t_{\mathrm{GUT}}) \approx 10^{82}\,\mathrm{m}^{-3}$, then after 100 e-foldings of inflation, the number density would have been $n_{\mathrm{M}}(t_f) = e^{-300} n_{\mathrm{M}}(t_{\mathrm{GUT}}) \approx 5 \times 10^{-49}\,\mathrm{m}^{-3} \approx 15\,\mathrm{pc}^{-3}$. The number density today, after the additional expansion from $a(t_f) \approx 2 \times 10^{-27}$ to $a_0 = 1$, would then be $n_{\mathrm{M}}(t_0) \approx 1 \times 10^{-61}\,\mathrm{Mpc}^{-3}$. The probability of finding even a single monopole within the last scattering surface would be astronomically small.

11.5 ■ THE PHYSICS OF INFLATION

Inflation explains some otherwise puzzling aspects of our universe, by flattening it, ensuring its homogeneity over large scales, and driving down the number density of magnetic monopoles it contains. However, we have not yet answered many crucial questions about the inflationary epoch. What triggers inflation at $t = t_i$, and (just as important) what turns it off at $t = t_f$? If inflation reduces the number density of monopoles to undetectably low levels, why doesn't it reduce the number density of photons to undetectably low levels? If inflation is so efficient at flattening the global curvature of the universe, why doesn't it also flatten out the local curvature due to fluctuations in the energy density? We know that the universe wasn't *perfectly* homogeneous after inflation, because the Cosmic Microwave Background isn't perfectly isotropic.

To answer these questions, we will have to examine, at least in broad outline, the physics behind inflation. At present, there is not a consensus among cosmologists about the exact mechanism driving inflation. We will restrict ourselves to speaking in general terms about one plausible mechanism for bringing about an inflationary epoch.

Suppose the universe contains a scalar field $\phi(\vec{r}, t)$ whose value can vary as a function of position and time. [Some early implementations of inflation associated the scalar field ϕ with the Higgs field, which mediates interactions between particles at energies higher than the GUT energy; however, to keep the discussion general, let's just call the field $\phi(\vec{r}, t)$ the *inflaton field*.] Generally speaking, a scalar field can have an associated potential energy $V(\phi)$.[2]

If ϕ has units of energy, and its potential V has units of energy density, then the energy density of the inflaton field is

$$\varepsilon_\phi = \frac{1}{2}\frac{1}{\hbar c^3}\dot{\phi}^2 + V(\phi) \tag{11.42}$$

in a region of space where ϕ is homogeneous. The pressure of the inflaton field is

$$P_\phi = \frac{1}{2}\frac{1}{\hbar c^3}\dot{\phi}^2 - V(\phi). \tag{11.43}$$

If the inflaton field changes only very slowly as a function of time, with

$$\dot{\phi}^2 \ll \hbar c^3 V(\phi), \tag{11.44}$$

then the inflaton field acts like a cosmological constant, with

$$\varepsilon_\phi \approx -P_\phi \approx V(\phi). \tag{11.45}$$

Thus, an inflaton field can drive exponential inflation if there is a temporary period when its rate of change $\dot{\phi}$ is small (satisfying equation (11.44)), and its potential $V(\phi)$ is large enough to dominate the energy density of the universe.

Under what circumstances are the conditions for inflation (small $\dot{\phi}$ and large V) met in the early universe? To determine the value of $\dot{\phi}$, start with the fluid equation for the energy density of the inflaton field,

$$\dot{\varepsilon}_\phi + 3H(t)(\varepsilon_\phi + P_\phi) = 0, \tag{11.46}$$

where $H(t) = \dot{a}/a$. Substituting from equations (11.42) and (11.43), we find the equation that governs the rate of change of ϕ:

$$\ddot{\phi} + 3H(t)\dot{\phi} = -\hbar c^3 \frac{dV}{d\phi}. \tag{11.47}$$

Note that equation (11.47) mimics the equation of motion for a particle being accelerated by a force proportional to $-dV/d\phi$ and being impeded by a frictional force proportional to the particle's speed. Thus, the expansion of the universe provides a "Hubble friction" term, $3H\dot{\phi}$, which slows the transition of the inflaton field to a value that will minimize the potential V. Just as a skydiver reaches terminal velocity when the downward force of gravity is balanced by the upward force of air resistance, so the inflaton field can reach "terminal velocity" (with $\ddot{\phi} = 0$) when

$$3H\dot{\phi} = -\hbar c^3 \frac{dV}{d\phi}, \tag{11.48}$$

[2]As a simple illustrative example, suppose that the scalar field ϕ is the elevation above sea level at a given point on the Earth's surface. The associated potential energy, in this case, is the gravitational potential $V = g\phi$, where $g = 9.8\,\mathrm{m\,s^{-2}}$.

or

$$\dot{\phi} = -\frac{\hbar c^3}{3H}\frac{dV}{d\phi}. \tag{11.49}$$

If the inflaton field has reached this terminal velocity, then the requirement that $\dot{\phi}^2 \ll \hbar c^3 V$, necessary if the inflaton field is to play the role of a cosmological constant, translates into

$$\left(\frac{dV}{d\phi}\right)^2 \ll \frac{9H^2 V}{\hbar c^3}. \tag{11.50}$$

If the universe is undergoing exponential inflation driven by the potential energy of the inflaton field, this means that the Hubble parameter is

$$H = \left(\frac{8\pi G\varepsilon_\phi}{3c^2}\right)^{1/2} = \left(\frac{8\pi G V}{3c^2}\right)^{1/2}, \tag{11.51}$$

and equation (11.50) becomes

$$\left(\frac{dV}{d\phi}\right)^2 \ll \frac{24\pi G V^2}{\hbar c^5}, \tag{11.52}$$

which can also be written as

$$\left(\frac{E_P}{V}\frac{dV}{d\phi}\right)^2 \ll 1, \tag{11.53}$$

where E_P is the Planck energy. If the slope of the inflaton's potential is sufficiently shallow, satisfying equation (11.53), and if the amplitude of the potential is sufficiently large to dominate the energy density of the universe, then the inflaton field is capable of giving rise to exponential expansion.

As a concrete example of a potential $V(\phi)$ which can give rise to inflation, consider the potential shown in Figure 11.3. The global minimum in the potential occurs when the value of the inflaton field is $\phi = \phi_0$. Suppose, however, that the inflaton field starts at $\phi \approx 0$, where the potential is $V(\phi) \approx V_0$. If

$$\left(\frac{dV}{d\phi}\right)^2 \ll \frac{V_0^2}{E_P^2} \tag{11.54}$$

on the "plateau" where $V \approx V_0$, then while ϕ is slowly rolling toward ϕ_0, the inflaton field contributes an energy density $\varepsilon_\phi \approx V_0 \approx$ constant to the universe.

When an inflaton field has a potential similar to that of Figure 11.3, it is referred to as being in a *metastable false vacuum state* when it is near the maximum at $\phi = 0$. Such a state is not truly stable; if the inflaton field is nudged from $\phi = 0$ to $\phi = +d\phi$, it will continue to slowly roll toward the *true vacuum* state at $\phi = \phi_0$ and $V = 0$. However, if the plateau is sufficiently broad as well as sufficiently

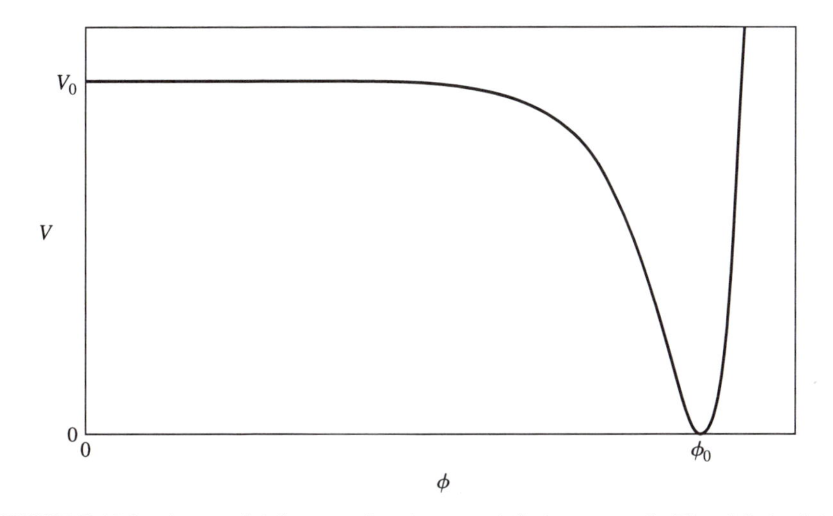

FIGURE 11.3 A potential that can give rise to an inflationary epoch. The global minimum in V (or "true vacuum") is at $\phi = \phi_0$. If the scalar field starts at $\phi = 0$, it is in a "false vacuum" state.

shallow, it can take many Hubble times for the inflaton field to roll down to the true vacuum state. Whether the inflaton field is dynamically significant during its transition from the false vacuum to the true vacuum depends on the value of V_0. As long as $\varepsilon_\phi \approx V_0$ is tiny compared to the energy density of radiation, $\varepsilon_r \sim \alpha T^4$, the contribution of the inflaton field to the Friedmann equation can be ignored. Exponential inflation, driven by the energy density of the inflaton field, will begin at a temperature

$$T_i \approx \left(\frac{V_0}{\alpha}\right)^{1/4} \approx 2 \times 10^{28}\,\mathrm{K} \left(\frac{V_0}{10^{105}\,\mathrm{TeV\,m^{-3}}}\right)^{1/4} \tag{11.55}$$

or

$$kT_i \approx (\hbar^3 c^3 V_0)^{1/4} \approx 2 \times 10^{12}\,\mathrm{TeV} \left(\frac{V_0}{10^{105}\,\mathrm{TeV\,m^{-3}}}\right)^{1/4}. \tag{11.56}$$

This corresponds to a time

$$t_i \approx \left(\frac{c^2}{GV_0}\right)^{1/2} \approx 3 \times 10^{-36}\,\mathrm{s} \left(\frac{V_0}{10^{105}\,\mathrm{TeV\,m^{-3}}}\right)^{-1/2}. \tag{11.57}$$

While the inflaton field is slowly rolling toward the true vacuum state, it produces exponential expansion, with a Hubble parameter

$$H_i \approx \left(\frac{8\pi G V_0}{3c^2}\right)^{1/2} \approx t_i^{-1}. \tag{11.58}$$

The exponential expansion ends as the inflaton field reaches the true vacuum at $\phi = \phi_0$. The duration of inflation thus depends on the exact shape of the potential $V(\phi)$. The number of e-foldings of inflation, for the potential shown in Figure 11.3, should be

$$N \sim H_i \frac{\phi_0}{\dot{\phi}} \sim \left(\frac{E_P}{V_0} \frac{dV}{d\phi} \right)^{-1} \left(\frac{\phi_0}{E_P} \right). \tag{11.59}$$

Large values of ϕ_0 and V_0 (that is, a broad, high plateau) and small values of $dV/d\phi$ (that is, a shallowly sloped plateau) lead to more e-foldings of inflation.

After rolling off the plateau in Figure 11.3, the inflaton field ϕ oscillates about the minimum at ϕ_0. The amplitude of these oscillations is damped by the "Hubble friction" term proportional to $H\dot{\phi}$ in equation (11.47). If the inflaton field is coupled to any of the other fields in the universe, however, the oscillations in ϕ are damped more rapidly, with the energy of the inflaton field being carried away by photons or other relativistic particles. These photons *reheat* the universe after the precipitous drop in temperature caused by inflation. The energy lost by the inflaton field after its phase transition from the false vacuum to the true vacuum can be thought of as the latent heat of that transition. When water freezes, to use a low-energy analogy, it loses an energy of $3 \times 10^8 \, \text{J} \, \text{m}^{-3}$, which goes to heat its surroundings.[3] Similarly, the transition from false to true vacuum releases an energy V_0, which goes to reheat the universe.

If the scale factor increases by a factor

$$\frac{a(t_f)}{a(t_i)} = e^N \tag{11.60}$$

during inflation, then the temperature will drop by a factor e^{-N}. If inflation starts around the GUT time, and lasts for $N = 100$ e-foldings, then the temperature drops from a toasty $T(t_i) \sim T_{\text{GUT}} \sim 10^{28}$ K to a chilly $T(t_f) \sim e^{-100} T_{\text{GUT}} \sim 10^{-15}$ K. At a temperature of 10^{-15} K, we'd expect to find a single photon in a box 25 AU on a side, as compared to the 411 million photons packed into every cubic meter of space today. Not only is inflation very effective at driving down the number density of magnetic monopoles, it is also effective at driving down the number density of every other type of particle, including photons. The chilly post-inflationary period didn't last, though. As the energy density associated with the inflaton field was converted to relativistic particles such as photons, the temperature of the universe was restored to its pre-inflationary value T_i.

Inflation successfully explains the flatness, homogeneity, and isotropy of the universe. It ensures that we live in a universe with a negligibly low density of magnetic monopoles, while the inclusion of reheating ensures that we *don't* live in a universe with a negligibly low density of photons. In some ways, though, infla-

[3]This is why orange growers spray their trees with water when a hard freeze threatens. The energy released by water as it freezes keeps the delicate leaves warm. (The thin layer of ice also cuts down on convective and radiative heat loss, but the release of latent heat is the largest effect.)

tion seems to be too successful. It makes the universe homogeneous and isotropic all right, but it makes it *too* homogeneous and isotropic. One hundred e-foldings of inflation not only flattens the global curvature of the universe, it also flattens the local curvature due to fluctuations in the energy density. If energy fluctuations prior to inflation were $\delta\varepsilon/\bar{\varepsilon} \sim 1$, a naïve calculation predicts that density fluctuations immediately after 100 e-foldings of inflation would be

$$\frac{\delta\varepsilon}{\bar{\varepsilon}} \sim e^{-100} \sim 10^{-43}. \tag{11.61}$$

This is a very close approach to homogeneity. Even allowing for the growth in amplitude of density fluctuations prior to the time of last scattering, this would leave the Cosmic Microwave Background much smoother than is actually observed.

Remember, however, the saga of how a submicroscopic patch of the universe ($d \sim 3 \times 10^{-44}$ m) was inflated to macroscopic size ($d \sim 1$ m), before growing to the size of the currently visible universe. Inflation excels in taking submicroscopic scales and blowing them up to macroscopic scales. On submicroscopic scales, the vacuum, whether true or false, is full of constantly changing quantum fluctuations, with virtual particles popping into and out of existence. On quantum scales, the universe is intrinsically inhomogeneous. Inflation takes the submicroscopic quantum fluctuations in the inflaton field and expands them to macroscopic scales. The energy fluctuations that result are the origin, in the inflationary scenario, of the inhomogeneities in the current universe. We can replace the old proverb, "Great oaks from tiny acorns grow," with the yet more amazing proverb, "Great superclusters from tiny quantum fluctuations grow."

SUGGESTED READING

Full references are given in the Annotated Bibliography on page 235.

Islam (2002), ch. 9: A general overview of inflation, avoiding technical concepts of particle physics

Liddle (1999), ch. 11: A brief, clear discussion of how inflation solves the horizon, flatness, and monopole problems

Liddle & Lyth (2000): A thorough, recent review of all aspects of inflationary cosmology

PROBLEMS

11.1. What upper limit is placed on $\Omega(t_P)$ by the requirement that the universe not end in a Big Crunch between the Planck time, $t_P \approx 5 \times 10^{-44}$ s, and the start of the inflationary epoch at t_i? Compute the maximum permissible value of $\Omega(t_P)$, first assuming $t_i \approx 10^{-36}$ s, then assuming $t_i \approx 10^{-26}$ s. (Hint: prior to inflation, the Friedmann equation will be dominated by the radiation term and the curvature term.)

11.2. Current observational limits on the density of magnetic monopoles tell us that their density parameter is currently $\Omega_{M,0} < 10^{-6}$. If monopoles formed at the GUT time, with one monopole per horizon of mass $m_M = m_{\text{GUT}}$, how many e-foldings of inflation would be required to drive the current density of monopoles below the bound $\Omega_{M,0} < 10^{-6}$? Assume that inflation took place immediately after the formation of monopoles.

11.3. It has been speculated that the present-day acceleration of the universe is due to the existence of a false vacuum, which will eventually decay. Suppose that the energy density of the false vacuum is $\varepsilon_\Lambda = 0.7\varepsilon_{c,0} = 3600\,\text{MeV}\,\text{m}^{-3}$, and that the current energy density of matter is $\varepsilon_{m,0} = 0.3\varepsilon_{c,0} = 1600\,\text{MeV}\,\text{m}^{-3}$. What will be the value of the Hubble parameter once the false vacuum becomes strongly dominant? Suppose that the false vacuum is fated to decay instantaneously to radiation at a time $t_f = 50t_0$. (Assume, for simplicity, that the radiation takes the form of blackbody photons.) To what temperature will the universe be reheated at $t = t_f$? What will the energy density of matter be at $t = t_f$? At what time will the universe again be dominated by matter?

CHAPTER 12

The Formation of Structure

The universe can be approximated as being homogeneous and isotropic only if we smooth it with a filter $\sim 100\,\text{Mpc}$ across. On smaller scales, the universe contains density fluctuations ranging from subatomic quantum fluctuations up to the large superclusters and voids, $\sim 50\,\text{Mpc}$ across, which characterize the distribution of galaxies in space. If we relax the strict assumption of homogeneity and isotropy that underlies the Robertson–Walker metric and the Friedmann equation, we can ask (and, to some extent, answer) the question, "How do density fluctuations in the universe evolve with time?"

The formation of relatively small objects, such as planets, stars, or even galaxies, involves some fairly complicated physics. Consider a galaxy, for instance. As mentioned in Chapter 8, the luminous portions of galaxies are typically much smaller than the dark halos in which they are embedded. In the usual scenario for galaxy formation, this is because the baryonic component of a galaxy radiates away energy, in the form of photons, and slides to the bottom of the potential well defined by the dark matter. The baryonic gas then fragments to form stars, in a nonlinear magnetohydrodynamical process.

In this chapter, however, we will focus on the formation of structures larger than galaxies—clusters, superclusters, and voids. Cosmologists use the term "large scale structure of the universe" to refer to all structures bigger than individual galaxies. A map of the large scale structure of the universe, as traced by the positions of galaxies, can be made by measuring the redshifts of a sample of galaxies and using the Hubble relation, $d = (c/H_0)z$, to compute their distances from our own galaxy. For instance, Figure 12.1 shows a redshift map from the 2dF Galaxy Redshift Survey. By plotting redshift as a function of angular position for galaxies in a long, narrow strip of the sky, a "slice of the universe" can be mapped. In a slice such as that of Figure 12.1, which reaches to $z \approx 0.3$, or $d_p(t_0) \approx 1300\,\text{Mpc}$, the galaxies obviously do not have a random Poisson distribution. The most prominent structures in Figure 12.1 are superclusters and voids. Superclusters are objects that are just in the process of collapsing under their own self-gravity. Superclusters are typically flattened (roughly planar) or elongated (roughly linear) structures. A supercluster will contain one or more clusters embedded within it; a cluster is a fully collapsed object that has come to equilibrium (more or less), and hence obeys the virial theorem, as discussed in section 8.3. In comparison to the flattened or elongated superclusters, the underdense voids

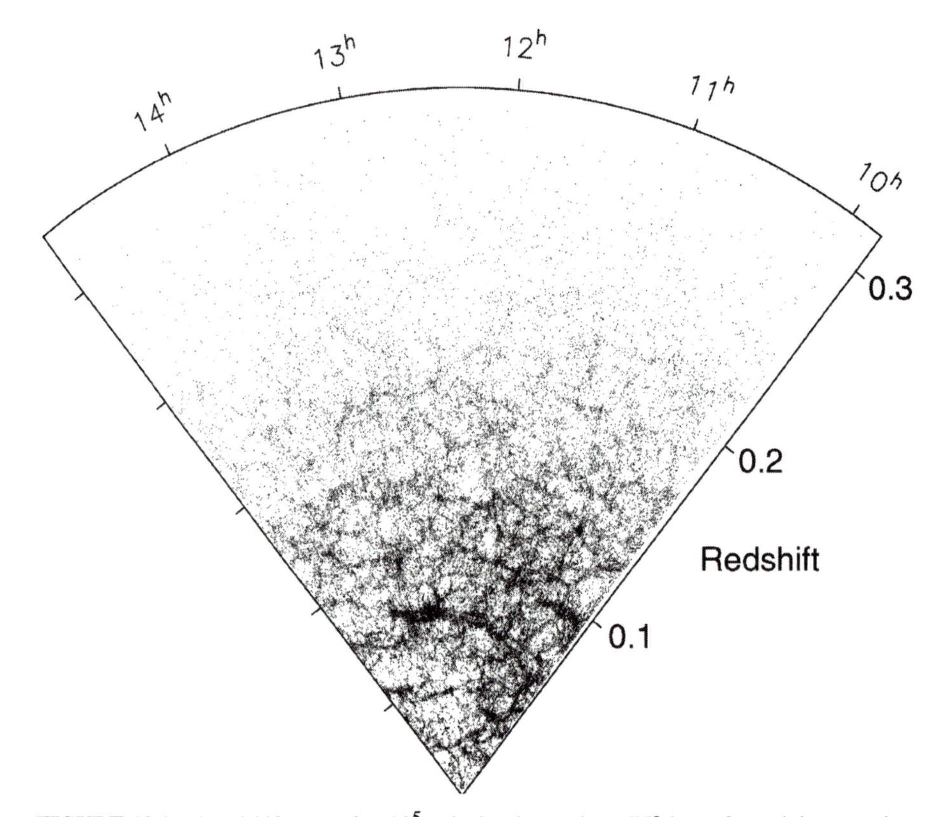

FIGURE 12.1 A redshift map of $\sim 10^5$ galaxies, in a strip $\sim 75°$ long, from right ascension $\alpha \approx 10\,\text{h}$ to $\alpha \approx 15\,\text{h}$, and $\sim 8°$ wide, from declination $\delta \approx -5°$ to $\delta \approx 3°$. (Image courtesy of the 2dF Galaxy Redshift Survey team.)

are roughly spherical in shape. When gazing at the large scale structure of the universe, as traced by the distribution of galaxies, cosmologists are likely to call it "bubbly" or "spongy" or "frothy" or "foamy."

Being able to describe the distribution of galaxies in space doesn't automatically lead to an understanding of the origin of large scale structure. Consider, as an analogy, the distribution of luminous objects shown in Figure 12.2. The distribution of illuminated cities on the Earth's surface is obviously not random. There are "superclusters" of cities, such as the Boswash supercluster stretching from Boston to Washington. There are "voids" such as the Appalachian void. However, the influences that determine the exact location of cities are often far removed from fundamental physics.[1]

Fortunately, the distribution of galaxies in space is more closely tied to fundamental physics than is the distribution of cities on the Earth. The basic mechanism

[1]Consider, for instance, the complicated politics that went into determining the location of Washington, DC.

FIGURE 12.2 The northeastern United States and southeastern Canada at night, as seen by a satellite from the Defense Meteorological Satellite Program (DMSP).

for growing large structures, such as voids and superclusters, is *gravitational instability*. Suppose that at some time in the past, the density of the universe had slight inhomogeneities. We know, for instance, that such density fluctuations occurred at the time of last scattering, since they left their stamp on the Cosmic Microwave Background. When the universe is matter-dominated, the overdense regions expand less rapidly than the universe as a whole; if their density is sufficiently great, they will collapse and become gravitationally bound objects such as clusters. The dense clusters will, in addition, draw matter to themselves from the surrounding underdense regions. The effect of gravity on density fluctuations is sometimes referred to as the Matthew Effect: "For whosoever hath, to him shall be given, and he shall have more abundance; but whosoever hath not, from him shall be taken away even that he hath" (Matthew 13:12). In less biblical language, the rich get richer and the poor get poorer.

12.1 ■ GRAVITATIONAL INSTABILITY

To put our study of gravitational instability on a more quantitative basis, consider some component of the universe whose energy density $\varepsilon(\vec{r}, t)$ is a function of position as well as time. At a given time t, the spatially averaged energy density is

$$\bar{\varepsilon}(t) = \frac{1}{V} \int_V \varepsilon(\vec{r}, t) d^3 r. \tag{12.1}$$

To ensure that we have found the true average, the volume V over which we are averaging must be large compared to the size of the biggest structure in the universe. It is useful to define a dimensionless density fluctuation

$$\delta(\vec{r}, t) \equiv \frac{\varepsilon(\vec{r}, t) - \bar{\varepsilon}(t)}{\bar{\varepsilon}(t)}. \tag{12.2}$$

The value of δ is thus negative in underdense regions and positive in overdense regions. The minimum possible value of δ is $\delta = -1$, corresponding to $\varepsilon = 0$. In principle, there is no upper limit on δ. You, for instance, represent a region of space where the baryon density has $\delta \approx 2 \times 10^{30}$.

The study of how large scale structure evolves with time requires knowing how a small fluctuation in density, with $|\delta| \ll 1$, grows in amplitude under the influence of gravity. This problem is most tractable when $|\delta|$ remains very much smaller than one. In the limit that the amplitude of the fluctuations remains small, we can successfully use linear perturbation theory.

To get a feel for how small density contrasts grow with time, consider a particularly simple case. Start with a static, homogeneous, matter-only universe with uniform mass density $\bar{\rho}$. (At this point, we stumble over the inconvenient fact that there's no such thing as a static, homogeneous, matter-only universe. This is the awkward fact that inspired Einstein to introduce the cosmological constant. However, there are conditions under which we can consider some region of the universe to be approximately static and homogeneous. For instance, the air in a closed room is approximately static and homogeneous; it is stabilized by a pressure gradient with a scale length much greater than the height of the ceiling.) In a region of the universe that is *approximately* static and homogeneous, we add a small amount of mass within a sphere of radius R, as seen in Figure 12.3, so that the density within the sphere is $\bar{\rho}(1+\delta)$, with $\delta \ll 1$. If the density excess δ is uniform within the sphere, then the gravitational acceleration at the sphere's surface, due to the excess mass, will be

$$\ddot{R} = -\frac{G(\Delta M)}{R^2} = -\frac{G}{R^2}\left(\frac{4\pi}{3}R^3\bar{\rho}\delta\right), \tag{12.3}$$

or

$$\frac{\ddot{R}}{R} = -\frac{4\pi G\bar{\rho}}{3}\delta(t). \tag{12.4}$$

Thus, a mass excess ($\delta > 0$) will cause the sphere to collapse inward ($\ddot{R} < 0$).

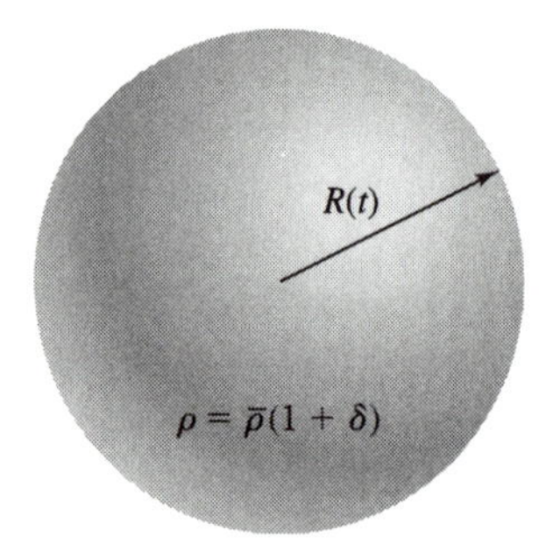

FIGURE 12.3 A sphere of radius $R(t)$ expanding or contracting under the influence of the density fluctuation $\delta(t)$.

Equation (12.4) contains two unknowns, $R(t)$ and $\delta(t)$. If we want to find an explicit solution for $\delta(t)$, we need a second equation involving $R(t)$ and $\delta(t)$. Conservation of mass tells us that the mass of the sphere,

$$M = \frac{4\pi}{3}\bar{\rho}[1 + \delta(t)]R(t)^3, \tag{12.5}$$

remains constant during the collapse. Thus we can write another relation between $R(t)$ and $\delta(t)$, which must hold true during the collapse:

$$R(t) = R_0[1 + \delta(t)]^{-1/3}, \tag{12.6}$$

where

$$R_0 \equiv \left(\frac{3M}{4\pi\bar{\rho}}\right)^{1/3} = \text{constant}. \tag{12.7}$$

When $\delta \ll 1$, we may make the approximation

$$R(t) \approx R_0\left[1 - \frac{1}{3}\delta(t)\right]. \tag{12.8}$$

Taking the second time derivative yields

$$\ddot{R} \approx -\frac{1}{3}R_0\ddot{\delta} \approx -\frac{1}{3}R\ddot{\delta}. \tag{12.9}$$

Thus, mass conservation tells us that

$$\frac{\ddot{R}}{R} \approx -\frac{1}{3}\ddot{\delta} \tag{12.10}$$

in the limit that $\delta \ll 1$. Combining equations (12.4) and (12.10), we find a tidy equation that tells us how the small overdensity δ evolves as the sphere collapses:

$$\ddot{\delta} = 4\pi G\bar{\rho}\delta. \tag{12.11}$$

The most general solution of equation (12.11) has the form

$$\delta(t) = A_1 e^{t/t_{\rm dyn}} + A_2 e^{-t/t_{\rm dyn}}, \tag{12.12}$$

where the dynamical time for collapse is

$$t_{\rm dyn} = \frac{1}{(4\pi G\bar{\rho})^{1/2}} = \left(\frac{c^2}{4\pi G\bar{\varepsilon}}\right)^{1/2}. \tag{12.13}$$

Note that the dynamical time depends only on $\bar{\rho}$, and not on R. The constants A_1 and A_2 in equation (12.12) depend on the initial conditions of the sphere. For instance, if the overdense sphere starts at rest, with $\dot{\delta} = 0$ at $t = 0$, then $A_1 = A_2 = \delta(0)/2$. After a few dynamical times, however, only the exponentially growing term of equation (12.12) is significant. Thus, gravity tends to make small density fluctuations in a static, pressureless medium grow exponentially with time.

12.2 ■ THE JEANS LENGTH

The exponential growth of density perturbations is slightly alarming, at first glance. For instance, the density of the air around you is $\bar{\rho} \approx 1\,\text{kg}\,\text{m}^{-3}$, yielding a dynamical time for collapse of $t_{\text{dyn}} \approx 9\,\text{hours}$.[2] What keeps small density perturbations in the air from undergoing a runaway collapse over the course of a few days? The answer, of course, is pressure. A nonrelativistic gas, as shown in section 4.3, has an equation-of-state parameter

$$w \approx \frac{kT}{\mu c^2}, \tag{12.14}$$

where T is the temperature of the gas and μ is the mean mass per gas particle. Thus, the pressure of an ideal gas will never totally vanish, but will only approach zero in the limit that the temperature approaches absolute zero.

When a sphere of gas is compressed by its own gravity, a pressure gradient will build up that tends to counter the effects of gravity.[3] However, hydrostatic equilibrium, the state in which gravity is exactly balanced by a pressure gradient, cannot always be attained. Consider an overdense sphere with initial radius R. If pressure were not present, it would collapse on a timescale

$$t_{\text{dyn}} \sim \frac{1}{(G\bar{\rho})^{1/2}} \sim \left(\frac{c^2}{G\bar{\varepsilon}}\right)^{1/2}. \tag{12.15}$$

If the pressure is nonzero, the attempted collapse will be countered by a steepening of the pressure gradient within the perturbation. The steepening of the pressure gradient, however, doesn't occur instantaneously. Any change in pressure travels at the sound speed.[4] Thus, the time it takes for the pressure gradient to build up in a region of radius R will be

$$t_{\text{pre}} \sim \frac{R}{c_s}, \tag{12.16}$$

where c_s is the local sound speed. In a medium with equation-of-state parameter $w > 0$, the sound speed is

$$c_s = c\left(\frac{dP}{d\varepsilon}\right)^{1/2} = \sqrt{w}c. \tag{12.17}$$

[2]Slightly longer if you are using this book for recreational reading as you climb Mount Everest.

[3]A star is the prime example of a dense sphere of gas in which the inward force of gravity is balanced by the outward force provided by a pressure gradient.

[4]What is sound, after all, but a traveling change in pressure?

For hydrostatic equilibrium to be attained, the pressure gradient must build up before the overdense region collapses, implying

$$t_{\rm pre} < t_{\rm dyn}. \tag{12.18}$$

Comparing equation (12.15) with equation (12.16), we find that for a density perturbation to be stabilized by pressure against collapse, it must be smaller than some reference size λ_J, given by the relation

$$\lambda_J \sim c_s t_{\rm dyn} \sim c_s \left(\frac{c^2}{G\bar{\varepsilon}} \right)^{1/2}. \tag{12.19}$$

The length scale λ_J is known as the *Jeans length*, after the astrophysicist James Jeans, who was among the first to study gravitational instability in a cosmological context. Overdense regions larger than the Jeans length collapse under their own gravity. Overdense regions smaller than the Jeans length merely oscillate in density; they constitute stable sound waves.

A more precise derivation of the Jeans length, including all the appropriate factors of π, yields the result

$$\lambda_J = c_s \left(\frac{\pi c^2}{G\bar{\varepsilon}} \right)^{1/2} = 2\pi c_s t_{\rm dyn}. \tag{12.20}$$

The Jeans length of the Earth's atmosphere, for instance, where the sound speed is a third of a kilometer per second and the dynamical time is nine hours, is $\lambda_J \sim 10^5$ km, far longer than the scale height of the Earth's atmosphere. You don't have to worry about density fluctuations in the air undergoing a catastrophic collapse.

To consider the behavior of density fluctuations on cosmological scales, consider a spatially flat universe in which the mean density is $\bar{\varepsilon}$, but which contains density fluctuations with amplitude $|\delta| \ll 1$. The characteristic time for expansion of such a universe is the Hubble time,

$$H^{-1} = \left(\frac{3c^2}{8\pi G\bar{\varepsilon}} \right)^{1/2}. \tag{12.21}$$

Comparison of equation (12.13) with equation (12.21) reveals that the Hubble time is comparable in magnitude to the dynamical time $t_{\rm dyn}$ for the collapse of an overdense region:

$$H^{-1} = \left(\frac{3}{2} \right)^{1/2} t_{\rm dyn} \approx 1.22 t_{\rm dyn}. \tag{12.22}$$

The Jeans length in an expanding flat universe will then be

$$\lambda_J = 2\pi c_s t_{\rm dyn} = 2\pi \left(\frac{2}{3} \right)^{1/2} \frac{c_s}{H}. \tag{12.23}$$

If we focus on one particular component of the universe, with equation-of-state parameter w and sound speed $c_s = \sqrt{w}c$, the Jeans length for that component will be

$$\lambda_J = 2\pi \left(\frac{2}{3}\right)^{1/2} \sqrt{w}\frac{c}{H}. \tag{12.24}$$

Consider, for instance, the "radiation" component of the universe. With $w = \frac{1}{3}$, the sound speed in a gas of photons or other relativistic particles is

$$c_s = c/\sqrt{3} \approx 0.58c. \tag{12.25}$$

The Jeans length for radiation in an expanding universe is then

$$\lambda_J = \frac{2\pi\sqrt{2}}{3}\frac{c}{H} \approx 3.0\frac{c}{H}. \tag{12.26}$$

Density fluctuations in the radiative component will be pressure-supported if they are smaller than three times the Hubble distance. Although a universe containing nothing but radiation can have density perturbations smaller than $\lambda_J \sim 3c/H$, they will be stable sound waves, and will not collapse under their own gravity.

In order for a universe to have gravitationally collapsed structures much smaller than the Hubble distance, it must have a nonrelativistic component, with $\sqrt{w} \ll 1$. The gravitational collapse of the *baryonic* component of the universe is complicated by the fact that it was coupled to photons until a redshift $z_{\rm dec} \approx z_{\rm ls} \approx 1100$. In section 9.5, the Hubble distance at the time of last scattering (effectively equal to the time of decoupling) was shown to be $c/H(z_{\rm dec}) \approx 0.2\,{\rm Mpc}$. The energy density of baryons at decoupling was $\varepsilon_{\rm bary} \approx 2.8 \times 10^{11}\,{\rm MeV\,m^{-3}}$, corresponding to a mass density $\rho_{\rm bary} \approx 5.0 \times 10^{-19}\,{\rm kg\,m^{-3}}$, and the energy density of photons was $\varepsilon_\gamma \approx 3.8 \times 10^{11}\,{\rm MeV\,m^{-3}} \approx 1.4\varepsilon_{\rm bary}$.

Prior to decoupling, the photons, electrons, and baryons were all coupled together to form a single photon-baryon fluid. Since the photons were still dominant over the baryons at the time of decoupling, with $\varepsilon_\gamma > \varepsilon_{\rm bary}$, we can regard the baryons (with only mild exaggeration) as being a dynamically insignificant contaminant in the photon-baryon fluid. Just *before* decoupling, if we regard the baryons as a minor contaminant, the Jeans length of the photon-baryon fluid was roughly the same as the Jeans length of a pure photon gas:

$$\lambda_J(\text{before}) \approx 3c/H(z_{\rm dec}) \approx 0.6\,{\rm Mpc} \approx 1.9 \times 10^{22}\,{\rm m}. \tag{12.27}$$

The *baryonic Jeans mass*, M_J, is defined as the mass of baryons contained within a sphere of radius λ_J:

$$M_J \equiv \rho_{\rm bary}\left(\frac{4\pi}{3}\lambda_J^3\right). \tag{12.28}$$

Immediately before decoupling, the baryonic Jeans mass was

$$
\begin{aligned}
M_J(\text{before}) &\approx 5.0 \times 10^{-19}\,\text{kg}\,\text{m}^{-3}\left(\frac{4\pi}{3}\right)(1.9 \times 10^{22}\,\text{m})^3 \\
&\approx 1.3 \times 10^{49}\,\text{kg} \approx 7 \times 10^{18}\,\text{M}_\odot.
\end{aligned} \tag{12.29}
$$

This is approximately 3×10^4 times greater than the estimated baryonic mass of the Coma cluster, and represents a mass greater than the baryonic mass of even the largest supercluster seen today.

Now consider what happens to the baryonic Jeans mass immediately after decoupling. Once the photons are decoupled, the photons and baryons form two separate gases, instead of a single photon-baryon fluid. The sound speed in the photon gas is

$$
c_s(\text{photon}) = c/\sqrt{3} \approx 0.58c. \tag{12.30}
$$

The sound speed in the baryonic gas, by contrast, is

$$
c_s(\text{baryon}) = \left(\frac{kT}{mc^2}\right)^{1/2} c. \tag{12.31}
$$

At the time of decoupling, the thermal energy per particle was $kT_{\text{dec}} \approx 0.26\,\text{eV}$, and the mean rest energy of the atoms in the baryonic gas was $mc^2 = 1.22 m_p c^2 \approx 1140\,\text{MeV}$, taking into account the helium mass fraction of $Y_p = 0.24$. Thus, the sound speed of the baryonic gas immediately after decoupling was

$$
c_s(\text{baryon}) \approx \left(\frac{0.26\,\text{eV}}{1140 \times 10^6\,\text{eV}}\right)^{1/2} c \approx 1.5 \times 10^{-5} c, \tag{12.32}
$$

only 5 kilometers per second. Thus, once the baryons were decoupled from the photons, their associated Jeans length decreased by a factor

$$
F = \frac{c_s(\text{baryon})}{c_s(\text{photon})} \approx \frac{1.5 \times 10^{-5}}{0.58} \approx 2.6 \times 10^{-5}. \tag{12.33}
$$

Decoupling causes the baryonic Jeans mass to decrease by a factor $F^3 \approx 1.8 \times 10^{-14}$, plummeting from $M_J(\text{before}) \approx 7 \times 10^{18}\,\text{M}_\odot$ to

$$
M_J(\text{after}) = F^3 M_J(\text{before}) \approx 1 \times 10^5\,\text{M}_\odot. \tag{12.34}
$$

This is comparable to the baryonic mass of the smallest dwarf galaxies known, and is very much smaller than the baryonic mass of our own galaxy, which is $\sim 10^{11}\,\text{M}_\odot$.

The abrupt decrease of the baryonic Jeans mass at the time of decoupling marks an important epoch in the history of structure formation. Perturbations in the baryon density, from supercluster scales down to the size of the smallest dwarf galaxies, couldn't grow in amplitude until the time of photon decoupling, when the universe had reached the ripe old age of $t_{\text{dec}} \approx 0.35\,\text{Myr}$. After decoupling, the growth of density perturbations in the baryonic component was off and running. The baryonic Jeans mass, already small by cosmological standards at the time of decoupling, dropped still further with time as the universe expanded and the baryonic component cooled.

12.3 ■ INSTABILITY IN AN EXPANDING UNIVERSE

Density perturbations smaller than the Hubble distance can grow in amplitude only when they are no longer pressure-supported. For the baryonic matter, this loss of pressure support happens abruptly at the time of decoupling, when the Jeans length for baryons drops suddenly by a factor $F \sim 3 \times 10^{-5}$. For the *dark* matter, the loss of pressure support occurs more gradually, as the thermal energy of the dark matter particles drops below their rest energy. When considering the Cosmic Neutrino Background, for instance, which has a temperature comparable to the Cosmic Microwave Background, we found (see equation (5.18)) that neutrinos of mass m_ν became nonrelativistic at a redshift

$$1 + z = \frac{1}{a} \approx \frac{m_\nu c^2}{5 \times 10^{-4}\,\text{eV}}. \tag{12.35}$$

Thus, if the universe contains a Cosmic WIMP Background comparable in temperature to the Cosmic Neutrino Background, the WIMPs, if they have a mass $m_W c^2 \gg 2\,\text{eV}$, would have become nonrelativistic long before the time of radiation-matter equality at $z_{rm} \approx 3570$.

Once the pressure (and hence the Jeans length) of some component becomes negligibly small, does this imply that the amplitude of density fluctuations is free to grow exponentially with time? Not necessarily. The analysis of section 12.1, which yielded $\delta \propto \exp(t/t_{dyn})$, assumed that the universe was *static* as well as pressureless. In an expanding Friedmann universe, the timescale for the growth of a density perturbation by self-gravity, $t_{dyn} \sim (c^2/G\bar{\varepsilon})^{1/2}$, is comparable to the timescale for expansion, $H^{-1} \sim (c^2/G\bar{\varepsilon})^{1/2}$. Self-gravity, in the absence of global expansion, causes overdense regions to become *more dense* with time. The global expansion of the universe, in the absence of self-gravity, causes overdense regions to become *less dense* with time. Because the timescales for these two competing processes are similar, they must both be taken into account when computing the time evolution of a density perturbation.

To get a feel how small density perturbations in an expanding universe evolve with time, let's do a Newtonian analysis of the problem, similar in spirit to the Newtonian derivation of the Friedmann equation given in Chapter 4. Suppose you are in a universe filled with pressureless matter with mass density $\bar{\rho}(t)$. As the universe expands, the density decreases at the rate $\bar{\rho}(t) \propto a(t)^{-3}$. Within a spherical region of radius R, a small amount of matter is added, or removed, so that the density within the sphere is

$$\rho(t) = \bar{\rho}(t)[1 + \delta(t)], \tag{12.36}$$

with $|\delta| \ll 1$. (In performing a Newtonian analysis of this problem, we are implicitly assuming that the radius R is small compared to the Hubble distance and large compared to the Jeans length.) The total gravitational acceleration at the surface of the sphere will be

$$\ddot{R} = -\frac{GM}{R^2} = -\frac{G}{R^2}\left(\frac{4\pi}{3}\rho R^3\right) = -\frac{4\pi}{3}G\bar{\rho}R - \frac{4\pi}{3}G(\bar{\rho}\delta)R. \tag{12.37}$$

The equation of motion for a point at the surface of the sphere can then be written in the form

$$\frac{\ddot{R}}{R} = -\frac{4\pi}{3}G\bar{\rho} - \frac{4\pi}{3}G\bar{\rho}\delta. \tag{12.38}$$

Mass conservation tells us that the mass inside the sphere,

$$M = \frac{4\pi}{3}\bar{\rho}(t)[1+\delta(t)]R(t)^3, \tag{12.39}$$

remains constant as the sphere expands. Thus,

$$R(t) \propto \bar{\rho}(t)^{-1/3}[1+\delta(t)]^{-1/3}, \tag{12.40}$$

or, since $\bar{\rho} \propto a^{-3}$,

$$R(t) \propto a(t)[1+\delta(t)]^{-1/3}. \tag{12.41}$$

That is, if the sphere is slightly overdense, its radius will grow slightly less rapidly than the scale factor $a(t)$. If the sphere is slightly underdense, it will grow slightly more rapidly than the scale factor.

Taking two time derivatives of equation (12.41) yields

$$\frac{\ddot{R}}{R} = \frac{\ddot{a}}{a} - \frac{1}{3}\ddot{\delta} - \frac{2}{3}\frac{\dot{a}}{a}\dot{\delta}, \tag{12.42}$$

when $|\delta| \ll 1$. Combining equations (12.38) and (12.42), we find

$$\frac{\ddot{a}}{a} - \frac{1}{3}\ddot{\delta} - \frac{2}{3}\frac{\dot{a}}{a}\dot{\delta} = -\frac{4\pi}{3}G\bar{\rho} - \frac{4\pi}{3}G\bar{\rho}\delta. \tag{12.43}$$

When $\delta = 0$, equation (12.43) reduces to

$$\frac{\ddot{a}}{a} = -\frac{4\pi}{3}G\bar{\rho}, \tag{12.44}$$

which is the correct acceleration equation for a homogeneous, isotropic universe containing nothing but pressureless matter (compare to equation (4.44)). By subtracting equation (12.44) from equation (12.43) to leave only the terms linear in the perturbation δ, we find the equation that governs the growth of small perturbations:

$$-\frac{1}{3}\ddot{\delta} - \frac{2}{3}\frac{\dot{a}}{a}\dot{\delta} = -\frac{4\pi}{3}G\bar{\rho}\delta, \tag{12.45}$$

or

$$\ddot{\delta} + 2H\dot{\delta} = 4\pi G\bar{\rho}\delta, \tag{12.46}$$

remembering that $H \equiv \dot{a}/a$. In a static universe, with $H = 0$, equation (12.46) reduces to the result we already found in equation (12.11):

$$\ddot{\delta} = 4\pi G\bar{\rho}\delta. \tag{12.47}$$

The additional term, $\propto H\dot{\delta}$, found in an expanding universe, is sometimes called the "Hubble friction" term; it acts to slow the growth of density perturbations in an expanding universe.

A fully relativistic calculation for the growth of density perturbations yields the more general result

$$\ddot{\delta} + 2H\dot{\delta} = \frac{4\pi G}{c^2}\bar{\varepsilon}_m\delta. \tag{12.48}$$

This form of the equation can be applied to a universe that contains components with nonnegligible pressure, such as radiation ($w = \frac{1}{3}$) or a cosmological constant ($w = -1$). In multiple-component universes, however, it should be remembered that δ represents the fluctuation in the density of *matter* alone. That is,

$$\delta = \frac{\varepsilon_m - \bar{\varepsilon}_m}{\bar{\varepsilon}_m}, \tag{12.49}$$

where $\bar{\varepsilon}_m(t)$, the average matter density, might be only a small fraction of $\bar{\varepsilon}(t)$, the average total density. Rewritten in terms of the density parameter for matter,

$$\Omega_m = \frac{\bar{\varepsilon}_m}{\varepsilon_c} = \frac{8\pi G\bar{\varepsilon}_m}{3c^2H^2}, \tag{12.50}$$

equation (12.48) takes the form

$$\ddot{\delta} + 2H\dot{\delta} - \frac{3}{2}\Omega_m H^2\delta = 0. \tag{12.51}$$

During epochs when the universe is not dominated by matter, density perturbations in the matter do not grow rapidly in amplitude. Take, for instance, the early radiation-dominated phase of the universe. During this epoch, $\Omega_m \ll 1$ and $H = 1/(2t)$, meaning that equation (12.51) takes the form

$$\ddot{\delta} + \frac{1}{t}\dot{\delta} \approx 0, \tag{12.52}$$

which has a solution of the form

$$\delta(t) \approx B_1 + B_2 \ln t. \tag{12.53}$$

During the radiation-dominated epoch, density fluctuations in the dark matter grew only at a logarithmic rate. In the far future, if the universe is indeed dominated by a cosmological constant, the density parameter for matter will again be negligibly small, the Hubble parameter will have the constant value $H = H_\Lambda$, and equation (12.51) will take the form

$$\ddot{\delta} + 2H_\Lambda \dot{\delta} \approx 0, \tag{12.54}$$

which has a solution of the form

$$\delta(t) \approx C_1 + C_2 e^{-2H_\Lambda t}. \tag{12.55}$$

In a lambda-dominated phase, therefore, fluctuations in the matter density reach a constant fractional amplitude, while the average matter density plummets at the rate $\bar{\varepsilon}_m \propto e^{-3H_\Lambda t}$.

Only when matter dominates the energy density can fluctuations in the matter density grow at a significant rate. In a flat, matter-dominated universe, $\Omega_m = 1$, $H = 2/(3t)$, and equation (12.51) takes the form

$$\ddot{\delta} + \frac{4}{3t}\dot{\delta} - \frac{2}{3t^2}\delta = 0. \tag{12.56}$$

If we guess that the solution to the above equation has the power-law form Dt^n, plugging this guess into the equation yields

$$n(n-1)Dt^{n-2} + \frac{4}{3t}nDt^{n-1} - \frac{2}{3t^2}Dt^n = 0, \tag{12.57}$$

or

$$n(n-1) + \frac{4}{3}n - \frac{2}{3} = 0. \tag{12.58}$$

The two possible solutions for this quadratic equation are $n = -1$ and $n = \frac{2}{3}$. Thus, the general solution for the time evolution of density perturbations in a spatially flat, matter-only universe is

$$\delta(t) \approx D_1 t^{2/3} + D_2 t^{-1}. \tag{12.59}$$

The values of D_1 and D_2 are determined by the initial conditions for $\delta(t)$. The decaying mode, $\propto t^{-1}$, eventually becomes negligibly small compared to the growing mode, $\propto t^{2/3}$. When the growing mode is the only survivor, the density perturbations in a flat, matter-only universe grow at the rate

$$\delta \propto t^{2/3} \propto a(t) \propto \frac{1}{1+z} \tag{12.60}$$

as long as $|\delta| \ll 1$.

When an overdense region attains an overdensity $\delta \sim 1$, its evolution can no longer be treated with a simple linear perturbation approach. Studies of the nonlinear evolution of structure are commonly made using numerical computer simulations, in which the matter filling the universe is modeled as a distribution of point masses interacting via Newtonian gravity. In these simulations, as in the real universe, when a region reaches an overdensity $\delta \sim 1$, it breaks away from the Hubble flow and collapses. After one or two oscillations in radius, the overdense region attains virial equilibrium as a gravitationally bound structure. If the baryonic matter within the structure is able to cool efficiently (by bremsstrahlung or some other process) it will radiate away energy and fall to the center. The centrally concentrated baryons can then proceed to form stars, becoming the visible portions of galaxies that we see today. The less concentrated nonbaryonic matter forms the dark halo within which the stellar component of the galaxy is embedded.

If baryonic matter were the only type of nonrelativistic matter in the universe, then density perturbations could have started to grow at $z_{\rm dec} \approx 1100$, and they could have grown in amplitude only by a factor ~ 1100 by the present day. However, the dominant form of nonrelativistic matter is dark matter. The density perturbations in the dark matter started to grow effectively at $z_{\rm rd} \approx 3570$. At the time of decoupling, the baryons fell into the preexisting gravitational wells of the dark matter. The situation is schematically illustrated in Figure 12.4. Having nonbaryonic dark matter allows the universe to get a "head start" on structure formation;

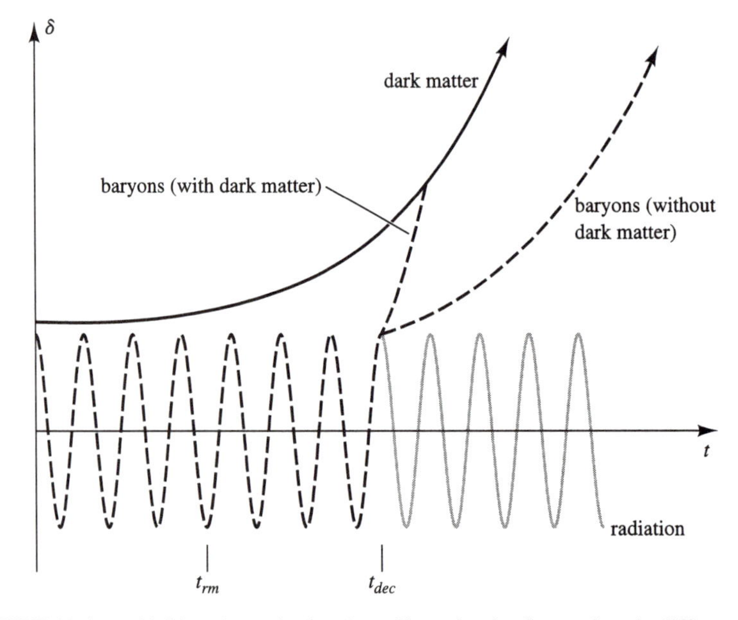

FIGURE 12.4 A highly schematic drawing of how density fluctuations in different components of the universe evolve with time.

perturbations in the matter density can start growing at $z_{\rm rm} \approx 3570$ rather than $z_{\rm dec} \approx 1100$, as they would in a universe without dark matter.

12.4 ■ THE POWER SPECTRUM

When deriving equation (12.46), which determines the growth rate of density perturbations in a Newtonian universe, we assumed that the perturbation was spherically symmetric. In fact, equation (12.46) and its relativistically correct brother, equation (12.48), both apply to low-amplitude perturbations of any shape. This is fortunate, since the density perturbations in the real universe are not spherically symmetric. The bubbly structure shown in redshift maps of galaxies, such as Figure 12.1, has grown from the density perturbations that were present when the universe became matter dominated. Great oaks from tiny acorns grow—but then, great pine trees from tiny pinenuts grow. By looking at the current large scale structure (the "tree"), we can deduce the properties of the early, low-amplitude, density fluctuations (the "nut").[5]

Let us consider the properties of the early density fluctuations at some time t_i, when they were still very low in amplitude ($|\delta| \ll 1$). As long as the density fluctuations are small in amplitude, the expansion of the universe is still nearly isotropic, and the geometry of the universe is still well described by the Robertson–Walker metric (equation (3.25)):

$$ds^2 = -c^2dt^2 + a(t)^2[dr^2 + S_\kappa(r)^2 d\Omega^2]. \qquad (12.61)$$

Under these circumstances, it is useful to set up a comoving coordinate system. Choose some point as the origin. In a universe described by the Robertson–Walker metric, as shown in section 3.4, the proper distance of any point from the origin can be written in the form

$$d_p(t_i) = a(t_i)r, \qquad (12.62)$$

where the comoving distance r is what the proper distance would be at the present day ($a = 1$) *if* the expansion continued to be perfectly isotropic. If we label each bit of matter in the universe with its comoving coordinate position $\vec{r} = (r, \theta, \phi)$, then $\vec{r}$ will remain very nearly constant as long as $|\delta| \ll 1$. Thus, when considering the regime where density fluctuations are small, cosmologists typically consider $\delta(\vec{r})$, the density fluctuation at a comoving location $\vec{r}$, at some time t_i. [The exact value of t_i doesn't matter, as long as it's a time after the density perturbations are in place, but before they reach an amplitude $|\delta| \sim 1$. Switching to a different value of t_i, under these restrictions, simply changes the amplitude of $\delta(\vec{r})$, and not its shape.]

When discussing the temperature fluctuations of the Cosmic Microwave Background in Chapter 9, I pointed out that cosmologists weren't interested in the

[5]At the risk of carrying the arboreal analogy too far, I should mention that the temperature fluctuations of the Cosmic Microwave Background, as shown in Figures 9.2 and 9.5, offer us a look at the "sapling."

exact pattern of hot and cold spots on the last scattering surface, but rather in the statistical properties of the field $\delta T/T(\theta, \phi)$. Similarly, cosmologists are not interested in the exact locations of the density maxima and minima in the early universe, but rather in the statistical properties of the field $\delta(\vec{r})$. When studying the temperature fluctuations of the CMB, it is useful to expand $\delta T/T(\phi, \theta)$ in spherical harmonics. A similar decomposition of $\delta(\vec{r})$ is also useful. Since δ is defined in three-dimensional space (rather than on the surface of a sphere), a useful expansion of δ is in terms of Fourier components.

Within a large comoving box, of comoving volume V, the density fluctuation field $\delta(\vec{r})$ can be expressed as

$$\delta(\vec{r}) = \frac{V}{(2\pi)^3} \int \delta_{\vec{k}} e^{-i\vec{k}\cdot\vec{r}} d^3k, \tag{12.63}$$

where the individual Fourier components $\delta_{\vec{k}}$ are found by performing the integral

$$\delta_{\vec{k}} = \frac{1}{V} \int \delta(\vec{r}) e^{i\vec{k}\cdot\vec{r}} d^3r. \tag{12.64}$$

In performing the Fourier transform, we are breaking up the function $\delta(\vec{r})$ into an infinite number of sine waves, each with comoving wavenumber $\vec{k}$ and comoving wavelength $\lambda = 2\pi/k$. If we have complete, uncensored knowledge of $\delta(\vec{r})$, we can compute all the Fourier components $\delta_{\vec{k}}$; conversely, if we know all the Fourier components, we can reconstruct the density field $\delta(\vec{r})$.

Each Fourier component is a complex number, which can be written in the form

$$\delta_{\vec{k}} = |\delta_{\vec{k}}| e^{i\phi_{\vec{k}}}. \tag{12.65}$$

When $|\delta_{\vec{k}}| \ll 1$, then each Fourier component obeys equation (12.51),

$$\ddot{\delta}_{\vec{k}} + 2H\dot{\delta}_{\vec{k}} - \frac{3}{2}\Omega_m H^2 \delta_{\vec{k}} = 0, \tag{12.66}$$

as long as the proper wavelength, $a(t)2\pi/k$, is large compared to the Jeans length and small compared to the Hubble distance c/H.[6] The phase $\phi_{\vec{k}}$ remains constant as long as the amplitude $|\delta_{\vec{k}}|$ remains small. Even after fluctuations with a short proper wavelength have reached $|\delta_{\vec{k}}| \sim 1$ and collapsed, the growth of the longer wavelength perturbations is still described by equation (12.66). This means, helpfully enough, that we can use linear perturbation theory to study the growth of very large scale structures even after smaller structures, such as galaxies and clusters of galaxies, have already collapsed.

[6] When a sine wave perturbation has a wavelength large compared to the Hubble distance, its crests are not causally connected to its troughs.

The mean square amplitude of the Fourier components defines the *power spectrum*:

$$P(k) = \langle |\delta_{\vec{k}}|^2 \rangle, \tag{12.67}$$

where the average is taken over all possible orientations of the wavenumber $\vec{k}$. [If $\delta(\vec{r})$ is isotropic, then no information is lost, statistically speaking, if we average the power spectrum over all angles.] When the phases $\phi_{\vec{k}}$ of the different Fourier components are uncorrelated with each other, then $\delta(\vec{r})$ is called a *Gaussian field*. If a Gaussian field is homogeneous and isotropic, then all its statistical properties are summed up in the power spectrum $P(k)$. If $\delta(\vec{r})$ is a Gaussian field, then the value of δ at a randomly selected point is drawn from the Gaussian probability distribution

$$p(\delta) = \frac{1}{\sqrt{2\pi}\sigma} \exp\left(-\frac{\delta^2}{2\sigma^2}\right), \tag{12.68}$$

where the standard deviation σ can be computed from the power spectrum:

$$\sigma = \frac{V}{(2\pi)^3} \int P(k) d^3k = \frac{V}{2\pi^2} \int_0^\infty P(k) k^2 dk. \tag{12.69}$$

The study of Gaussian density fields is of particular interest to cosmologists because most inflationary scenarios predict that the density fluctuations created by inflation (see section 11.5) will be an isotropic, homogeneous Gaussian field. In addition, the expected power spectrum for the inflationary fluctuations has a scale-invariant, power-law form:

$$P(k) \propto k^n, \tag{12.70}$$

with the favored value of the power-law index being $n = 1$. The preferred power spectrum, $P(k) \propto k$, is often referred to as a Harrison–Zel'dovich spectrum.

What would a universe with $P(k) \propto k^n$ look like? Imagine going through such a universe and marking out randomly located spheres of comoving radius L. The mean mass of each sphere (considering only the nonrelativistic matter which it contains) will be

$$\langle M \rangle = \frac{4\pi}{3} L^3 \frac{\varepsilon_{m,0}}{c^2}. \tag{12.71}$$

However, the actual mass of each sphere will vary; some spheres will be slightly underdense, and others will be slightly overdense. The mean square density fluctuation of the mass inside each sphere is a function of the power spectrum and of the size of the sphere:

$$\left\langle \left(\frac{M - \langle M \rangle}{\langle M \rangle} \right)^2 \right\rangle \propto k^3 P(k), \tag{12.72}$$

where the comoving wavenumber associated with the sphere is $k = 2\pi/L$. Thus, if the power spectrum has the form $P(k) \propto k^n$, the root mean square mass fluctuation within spheres of comoving radius L will be

$$\frac{\delta M}{M} \equiv \left\langle \left(\frac{M - \langle M \rangle}{\langle M \rangle} \right)^2 \right\rangle^{1/2} \propto L^{-(3+n)/2}. \tag{12.73}$$

This can also be expressed in the form $\delta M/M \propto M^{-(3+n)/6}$. For $n < -3$, the mass fluctuations diverge on large scales, which would be bad news for our assumption of homogeneity on large scales. (Note that if you scattered point masses randomly throughout the universe, so that they formed a Poisson distribution, you would expect mass fluctuations of amplitude $\delta M/M \propto N^{-1/2}$, where N is the expected number of point masses within the sphere. Since the average mass within a sphere is $M \propto N$, a Poisson point distribution has $\delta M/M \propto M^{-1/2}$, or $n = 0$. The Harrison–Zel'dovich spectrum, with $n = 1$, thus will produce more power on small length scales than a Poisson distribution of points.) Note that the potential fluctuations associated with mass fluctuations on a length scale L will have an amplitude $\delta\Phi \sim G\delta M/L \propto \delta M/M^{1/3} \propto M^{(1-n)/6}$. Thus, the Harrison–Zel'dovich spectrum, with $n = 1$, is the only power law that prevents the divergence of the potential fluctuations on both large and small scales.

12.5 ■ HOT VERSUS COLD

Immediately after inflation, the expected power spectrum for density perturbations has the form $P(k) \propto k^n$, with an index $n = 1$ being predicted by most inflationary models. However, the shape of the power spectrum will be modified between the end of inflation at t_f and the time of radiation-matter equality at $t_{\rm rm} \approx 4.7 \times 10^4$ yr. The shape of the power spectrum at $t_{\rm rm}$, when density perturbations start to grow significantly in amplitude, depends on the properties of the dark matter. More specifically, it depends on whether the dark matter is predominantly *cold dark matter* or *hot dark matter*.

Cold dark matter consists of particles that were nonrelativistic at the time they decoupled from the other components of the universe. For instance, WIMPs would have had thermal velocities much smaller than c at the time they decoupled, and hence qualify as cold dark matter. If any primordial black holes had formed in the early universe, their peculiar velocities would have been much smaller than c at the time they formed; thus primordial black holes would also act as cold dark matter. Axions are a type of elementary particle first proposed by particle physicists for noncosmological purposes. If they exist, however, they would have formed out of equilibrium in the early universe, with very low thermal velocities. Thus, axions would act as cold dark matter as well.

Hot dark matter, by contrast, consists of particles that were *relativistic* at the time they decoupled from the other components of the universe, and that remained relativistic until the mass contained within a Hubble volume (a sphere of proper radius c/H) was large compared to the mass of a galaxy. In the Benchmark Model, the Hubble distance at the time of radiation-matter equality was

$$\frac{c}{H(t_{\rm rm})} = \frac{c}{\sqrt{2}H_0} \frac{\Omega_{r,0}^{3/2}}{\Omega_{m,0}^2} \approx 1.8ct_{\rm rm} \approx 0.026\,{\rm Mpc}, \tag{12.74}$$

so the mass within a Hubble volume at that time was

$$\frac{4\pi}{3} \frac{c^3}{H(t_{\rm rm})^3} \frac{\Omega_{m,0}\rho_{c,0}}{a_{\rm rm}^3} = \frac{\sqrt{2}\pi}{3} \frac{c^3}{H_0^3} \frac{\Omega_{r,0}^{3/2}}{\Omega_{m,0}^2} \rho_{c,0} \approx 1.4 \times 10^{17}\,{\rm M}_\odot, \tag{12.75}$$

much larger than the mass of even a fairly large galaxy such as our own ($M_{\rm gal} \approx 10^{12}\,{\rm M}_\odot$). Thus, a weakly interacting particle that remains relativistic until the universe becomes matter-dominated will act as hot dark matter. For instance, neutrinos decoupled at $t \sim 1$ s, when the universe had a temperature $kT \sim 1$ MeV. Thus, a neutrino with mass $m_\nu c^2 \ll 1$ MeV was hot enough to be relativistic at the time it decoupled. Moreover, as discussed in section 5.1, a neutrino with mass $m_\nu c^2 < 2$ eV doesn't become nonrelativistic until after radiation-matter equality, and hence qualifies as hot dark matter.[7]

To see how the existence of hot dark matter modifies the spectrum of density perturbations, consider what would happen in a universe filled with weakly interacting particles, which are relativistic at the time they decouple. The initially relativistic particles cool as the universe expands, until their thermal velocities drop well below c when $3kT \sim m_h c^2$. This happens at a temperature

$$T_h \sim \frac{m_h c^2}{3k} \sim 8000\,{\rm K} \left(\frac{m_h c^2}{2\,{\rm eV}} \right). \tag{12.76}$$

In the radiation-dominated universe, this corresponds to a cosmic time (equation (10.2))

$$t_h \sim 2 \times 10^{12}\,{\rm s} \left(\frac{m_h c^2}{2\,{\rm eV}} \right)^{-2}. \tag{12.77}$$

Prior to the time t_h, the hot dark matter particles move freely in random directions with a speed close to that of light. This motion, called *free streaming*, acts to wipe out any density fluctuations present in the hot dark matter. Thus, the net effect of free streaming in the hot dark matter is to wipe out any density fluctuations whose wavelength is smaller than $\sim ct_h$. When the hot dark matter particles become nonrelativistic, there will be no density fluctuations on scales smaller than the

[7] It may seem odd to refer to neutrinos as "hot" dark matter, when the temperature of the Cosmic Neutrino Background is only two degrees above absolute zero. The label hot, in this case, simply means that the neutrinos were hot enough to be relativistic back in the radiation-dominated era.

physical scale

$$\lambda_{\min} \sim ct_h \sim 20\,\mathrm{kpc}\left(\frac{m_h c^2}{2\,\mathrm{eV}}\right)^{-2}, \tag{12.78}$$

corresponding to a comoving length scale

$$L_{\min} = \frac{\lambda_{\min}}{a(t_h)} \sim \frac{T_h}{2.725\,\mathrm{K}}\lambda_{\min} \sim 60\,\mathrm{Mpc}\left(\frac{m_h c^2}{2\,\mathrm{eV}}\right)^{-1}. \tag{12.79}$$

The total amount of matter within a sphere of comoving radius $L_{\min}$ is

$$M_{\min} = \frac{4\pi}{3}L_{\min}^3 \Omega_{m,0}\rho_{c,0} \sim 5 \times 10^{16}\,\mathrm{M}_\odot\left(\frac{m_h c^2}{2\,\mathrm{eV}}\right)^{-3}, \tag{12.80}$$

assuming $\Omega_{m,0} = 0.3$. If the dark matter is contributed by neutrinos with rest energy of a few electron volts, then the free streaming will wipe out all density fluctuations smaller than superclusters.

The upper panel of Figure 12.5 shows the power spectrum of density fluctuations in hot dark matter, once the hot dark matter has cooled enough to become nonrelativistic. Note that for wavenumbers $k \ll 2\pi/L_{\min}$, the power spectrum of hot dark matter (shown as the dotted line) is indistinguishable from the original $P \propto k$ spectrum (shown as the dashed line). However, the free streaming of the hot dark matter results in a severe loss of power for wavenumbers $k \gg 2\pi/L_{\min}$. The lower panel of Figure 12.5 shows that the root mean square mass fluctuations in hot dark matter, $\delta M/M \propto (k^3 P)^{1/2}$, have a maximum amplitude at a mass scale $M \sim 10^{16}\,\mathrm{M}_\odot$. This implies that in a universe filled with hot dark matter, the first structures to collapse are the size of superclusters. Smaller structures, such as clusters and galaxies then form by fragmentation of the superclusters. (This scenario, in which the largest observable structures form first, is called the *top-down* scenario.)

If most of the dark matter in the universe were hot dark matter, such as neutrinos, then we would expect the oldest structures in the universe to be superclusters, and that galaxies would be relatively young. In fact, the opposite seems to be true in our universe. Superclusters are just collapsing today, while galaxies have been around since at least $z \sim 6$, when the universe was less than a gigayear old. Thus, most of the dark matter in the universe must be *cold* dark matter, for which free streaming has been negligible.

The evolution of the power spectrum of cold dark matter, given the absence of free streaming, is quite different from the evolution of the power spectrum for hot dark matter. Remember, when the universe is radiation-dominated, density fluctuations $\delta_{\vec{k}}$ in the dark matter do not grow appreciably in amplitude, as long as their proper wavelength $a(t)2\pi/k$ is small compared to the Hubble distance $c/H(t)$. However, when the proper wavelength of a density perturbation is large

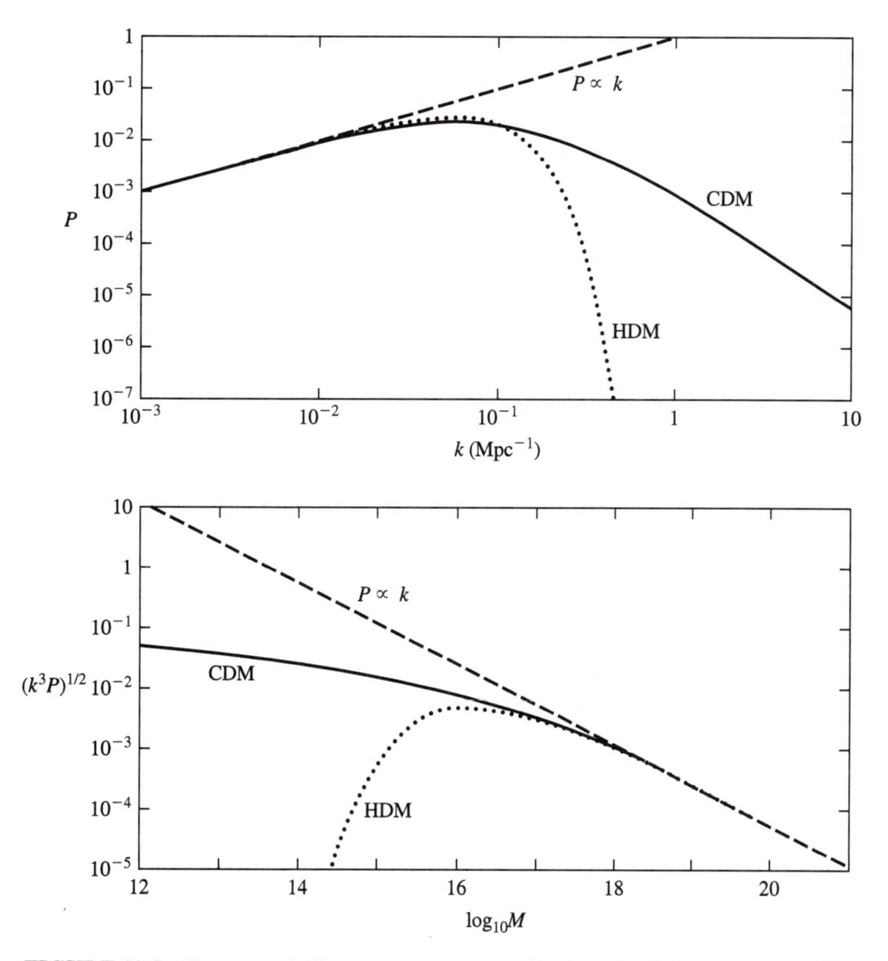

FIGURE 12.5 Upper panel: The power spectrum at the time of radiation-matter equality for cold dark matter (solid line) and for hot dark matter (dotted line). The initial power spectrum produced by inflation (dashed line) is assumed to have the form $P(k) \propto k$. The normalization of the power spectrum is arbitrary. Lower panel: The root mean square mass fluctuations, $\delta M/M \propto (k^3 P)^{1/2}$, are shown as a function of $M \propto k^{-3}$ (masses are in units of $M_\odot$). The line types are the same as in the upper panel.

compared to the Hubble distance, its amplitude will be able to increase, regardless of whether the universe is radiation-dominated or matter-dominated. If the cold dark matter consists of WIMPs, they decouple from the radiation at a time $t_d \sim 1$ s, when the scale factor is $a_d \sim 3 \times 10^{-10}$. At the time of WIMP decoupling, the Hubble distance is $c/H \sim 2ct_d \sim 6 \times 10^8$ m, corresponding to a comoving wavenumber

$$k_d \sim \frac{2\pi a_d}{2ct_d} \sim 10^5\,\mathrm{Mpc}^{-1}. \tag{12.81}$$

Thus, density fluctuations with a wavenumber $k < k_d$ will have a wavelength greater than the Hubble distance at the time of WIMP decoupling, and will be able to grow freely in amplitude, as long as their wavelength remains longer than the Hubble distance. Density fluctuations with $k > k_d$ will remain frozen in amplitude until matter starts to dominate the universe at $t_{\rm rm} \approx 4.7 \times 10^4$ yr, when the scale factor has grown to $a_{\rm rm} \approx 2.8 \times 10^{-4}$. At the time of radiation-matter equality, the Hubble distance, as given in equation (12.74), is $c/H \approx 1.8ct_{\rm rm} \approx 0.026\,{\rm Mpc}$, corresponding to a comoving wavenumber

$$k_{\rm rm} \approx \frac{2\pi a_{\rm rm}}{1.8ct_{\rm rm}} \approx 0.07\,{\rm Mpc}^{-1}. \tag{12.82}$$

Thus, density fluctuations with a wavenumber $k < k_{\rm rm} \approx 0.07\,{\rm Mpc}^{-1}$ will grow steadily in amplitude during the entire radiation-dominated era, and for wavenumbers $k < k_{\rm rm} \approx 0.07\,{\rm Mpc}^{-1}$, the power spectrum for cold dark matter retains the original $P(k) \propto k$ form, which it had immediately after inflation (see the upper panel of Figure 12.5).

By contrast, cold dark matter density perturbations with a wavenumber $k_d > k > k_{\rm rm}$ will be able to grow in amplitude only until their physical wavelength $a(t)/(2\pi k) \propto t^{1/2}$ is smaller than the Hubble distance $c/H(t) \propto t$. At that time, their amplitude will be frozen until the time $t_{\rm rm}$, when matter dominates, and density perturbations smaller than the Hubble distance are free to grow again. Thus, for wavenumbers $k > k_{\rm rm}$, the power spectrum for cold dark matter is suppressed in amplitude, with the suppression being greatest for the largest wavenumbers (corresponding to shorter wavelengths, which come within the horizon at an earlier time). The top panel of Figure 12.5 shows, as the solid line, the power spectrum for cold dark matter at the time of radiation-matter equality. Note the broad maximum in the power spectrum at $k \sim k_{\rm rm} \approx 0.07\,{\rm Mpc}^{-1}$. The root mean square mass fluctuations in the cold dark matter, shown in the bottom panel of Figure 12.5, are largest in amplitude for the smallest mass scales. This implies that in a universe filled with cold dark matter, the first objects to form are the *smallest*, with galaxies forming first, then clusters, then superclusters. This scenario, called the *bottom-up* scenario, is consistent with the observed ages of galaxies and superclusters.

Assuming that the dark matter consists of nothing but hot dark matter gives a poor fit to the observed large scale structure of the universe. Assuming that the dark matter is purely cold dark matter gives a much better fit. However, there is strong evidence that neutrinos do have some mass, and thus that the universe contains at least *some* hot dark matter. Cosmologists studying the large scale structure of the universe can adjust the assumed power spectrum of the dark matter, by mixing together hot and cold matter. (It's a bit like adjusting the temperature of your bath by tweaking the hot and cold water knobs.) Comparison of the assumed power spectrum to the observed large scale structure (as seen, for instance, in Figure 12.1) reveals that $\sim 13\%$ or less of the matter in the universe consists of hot dark matter. For $\Omega_{m,0} = 0.3$, this implies $\Omega_{\rm HDM,0} \leq 0.04$. If there were more hot dark matter than this amount, free streaming of the hot dark matter particles would make the universe too smooth on small scales. Some like it hot, but most like it cold—the majority of the dark matter in the universe must be *cold* dark matter.

SUGGESTED READING

Full references are given in the Annotated Bibliography on page 235.

Liddle & Lyth (2000): The origin of density perturbations during the inflationary era, and their growth thereafter

Longair (1998): For those who want to know more about galaxy formation, and how it ties into cosmology

Rich (2001), ch. 7: The origin and evolution of density fluctuations

PROBLEMS

12.1. Consider a spatially flat, matter-dominated universe ($\Omega = \Omega_m = 1$) that is *contracting* with time. What is the functional form of $\delta(t)$ in such a universe?

12.2. Consider an empty, negatively curved, expanding universe, as described in section 5.2. If a dynamically insignificant amount of matter ($\Omega_m \ll 1$) is present in such a universe, how do density fluctuations in the matter evolve with time? That is, what is the functional form of $\delta(t)$?

12.3. A volume containing a photon-baryon fluid is adiabatically expanded or compressed. The energy density of the fluid is $\varepsilon = \varepsilon_\gamma + \varepsilon_{\text{bary}}$, and the pressure is $P = P_\gamma = \varepsilon_\gamma/3$. What is $dP/d\varepsilon$ for the photon-baryon fluid? What is the sound speed, c_s? In equation (12.27), how large an error did we make in our estimate of λ_J(before) by ignoring the effect of the baryons on the sound speed of the photon-baryon fluid?

12.4. Suppose that the stars in a disk galaxy have a constant orbital speed v out to the edge of its spherical dark halo, at a distance R_{halo} from the galaxy's center. What is the average density $\bar{\rho}$ of the matter in the galaxy, including its dark halo? (Hint: go back to section 8.2.) What is the value of $\bar{\rho}$ for our galaxy, assuming $v = 220\,\text{km}\,\text{s}^{-1}$ and $R_{\text{halo}} = 100\,\text{kpc}$? If a bound structure, such as a galaxy, forms by gravitational collapse of an initially small density perturbation, the minimum time for collapse is $t_{\text{min}} \approx t_{\text{dyn}} \approx 1/\sqrt{G\bar{\rho}}$. Show that $t_{\text{min}} \approx R_{\text{halo}}/v$ for a disk galaxy. What is t_{min} for our own galaxy? What is the maximum possible redshift at which you would expect to see galaxies comparable in v and R_{halo} to our own galaxy? (Assume the Benchmark Model is correct.)

12.5. Within the Coma cluster, as discussed in section 8.3, galaxies have a root mean square velocity of $\langle v^2 \rangle^{1/2} \approx 1520\,\text{km}\,\text{s}^{-1}$ relative to the center of mass of the cluster; the half-mass radius of the Coma cluster is $r_h \approx 1.5\,\text{Mpc}$. Using arguments similar to those of the previous problem, compute the minimum time t_{min} required for the Coma cluster to form by gravitational collapse.

12.6. Derive equation (12.74), giving the Hubble distance at the time of radiation-matter equality. What was the Hubble distance at the time of matter-lambda equality, in the Benchmark Model? How much matter was contained within a Hubble volume at the time of matter-lambda equality?

12.7. Warm dark matter is defined as matter that became nonrelativistic when the amount of matter within a Hubble volume had a mass comparable to that of a galaxy. In the Benchmark Model, at what time $t_{\rm WDM}$ was the mass contained within a Hubble volume equal to $M_{\rm gal} = 10^{12}\,{\rm M}_\odot$? If the warm dark matter particles have a temperature equal to that of the cosmic neutrino background, what mass must they have in order to have become nonrelativistic at $t \sim t_{\rm WDM}$?

Epilogue

A book dealing with an active field like cosmology can't really have a neat, tidy ending. Our understanding of the universe is still growing and evolving. During the twentieth century, the growing weight of evidence pointed toward the Hot Big Bang model, in which the universe started in a hot, dense state, but gradually cooled as it expanded. At the end of the twentieth century and the beginning of the twenty-first, cosmological evidence was gathered at an increasing rate, refining our knowledge of the universe. As I write this epilogue, on a sunny spring day in the year 2002, the available evidence is explained by a Benchmark Model that is spatially flat and that has an expansion, which is currently accelerating. It seems that 70% of the energy density of the universe is contributed by a cosmological constant (or other form of "dark energy" with negative pressure). Only 30% of the energy density is contributed by matter (and only 4% is contributed by the familiar baryonic matter of which you and I are made).

However, many questions about the cosmos remain unanswered. Here are a few of the questions that currently nag at cosmologists:

- *What are the precise values of cosmological parameters such as H_0, q_0, $\Omega_{m,0}$, and $\Omega_{\Lambda,0}$?* Much effort has been invested in determining these parameters, but still they are not pinned down precisely.

- *What is the dark matter?* It can't be made entirely of baryons. It can't be made entirely of neutrinos. Most of the dark matter must be in the form of some exotic stuff that has not yet been detected in laboratories.

- *What is the dark energy?* Is it vacuum energy that plays the role of a cosmological constant, or is it some other component of the universe with $-1 < w < -\frac{1}{3}$? If it is vacuum energy, is it provided by a false vacuum, driving a temporary inflationary stage, or are we finally seeing the true vacuum energy?

- *What drove inflation during the early universe?* Our knowledge of the particle physics behind inflation is still sadly incomplete. Indeed, some cosmologists pose the questions, "Did inflation take place at all during the early universe? Is there another way to resolve the flatness, horizon, and monopole problems?"

- *Why is the universe expanding?* At one level, this question is easily answered. The universe is expanding today because it was expanding yesterday. It was expanding yesterday because it was expanding the day before yesterday … . However, when we extrapolate back to the Planck time, we find that the universe was expanding then with a Hubble parameter $H \sim 1/t_P$. What determined this set of initial conditions? In other words, "What put the Bang in the Big Bang?"

The most interesting questions, however, are those that we are still too ignorant to pose correctly. For instance, in ancient Egypt, a list of unanswered questions in cosmology might have included "How high is the dome that makes up the sky?" and "What's the dome made of?" Severely erroneous models of the universe obviously give rise to irrelevant questions. The exciting, unsettling possibility exists that future observations will render the now-promising Benchmark Model obsolete. I hope, patient reader, that learning about cosmology from this book has encouraged you to become a cosmologist yourself, and to join the scientists who are laboring to make my book a quaint, out-of-date relic from a time when the universe was poorly understood.

Annotated Bibliography

Works described as Popular contain little or no math. Those described as Intermediate are at roughly the same level as this book. Those described as Advanced have a higher level of mathematical and physical sophistication, appropriate for study at a graduate level.

POPULAR

Harrison, E. R., *Darkness at Night: A Riddle of the Universe* (Cambridge: Harvard University Press) 1987. A comprehensive discussion of Olbers' Paradox and its place in the history of cosmology.

Kragh, H., *Cosmology and Controversy* (Princeton: Princeton University Press) 1996. A well-reseached history of the Big Bang versus Steady State debate. A fascinating book if you are at all interested in the sociology of science.

Silk, J., *The Big Bang*, third edition (New York: W. H. Freeman & Co.) 2001. A broad overview of cosmology. Although aimed at a popular audience (with all mathematical formulas banished to an appendix), it doesn't skimp on the physics.

Weinberg, S., *The First Three Minutes*, revised edition (New York: Perseus Books) 1993. A classic of popular science. Weinberg's revision has brought the original 1977 version more nearly up-to-date.

INTERMEDIATE

Bernstein, J., *Introduction to Cosmology* (Englewood Cliffs, NJ: Prentice Hall) 1995. Has a slightly greater emphasis on particle physics than most cosmology texts.

Coles, P., *The Routledge Critical Dictionary of the New Cosmology* (New York: Routledge) 1999. In addition to a dictionary of cosmology-related terms, from "absorption line" to "Zel'dovich–Sunyaev effect," this book also contains longer essays on cosmological topics of current interest.

Cox, A. N., ed., *Allen's Astrophysical Quantities*, fourth edition (New York: Springer-Verlag) 2000. A standard reference book of astronomically relevant data, from the Euler vectors of the Nazca plate to the intensity of the extragalactic gamma-ray background.

Harrison, E., *Cosmology: The Science of the Universe*, second edition (Cambridge: Cambridge University Press) 2000. A wide-ranging book, placing the science of cosmology in its historical context, and discussing its philosophical and religious implications.

Islam, J. N., *An Introduction to Mathematical Cosmology*, second edition (Cambridge: Cambridge University Press) 2002. A book that emphasizes (as its name implies) the mathematical rather than the observational aspects of cosmology.

Liddle, A., *An Introduction to Modern Cosmology* (Chichester: John Wiley & Sons) 1999. A clear and concise introductory work.

Longair, Malcolm S., *Galaxy Formation* (Berlin: Springer-Verlag) 1998. A well-written introduction to galaxy formation, approached from a cosmological perspective.

Narlikar, J. V., *Introduction to Cosmology*, third edition (Cambridge: Cambridge University Press) 2002. Particularly useful for its section on alternative (non-Friedmann) cosmologies.

Peacock, J. A., *Cosmological Physics* (Cambridge: Cambridge University Press) 1999. A large, well-stuffed grabbag of cosmological topics. Contains, among other useful things, a long, detailed discussion of galaxy formation and clustering.

Rich, J., *Fundamentals of Cosmology* (Berlin: Springer-Verlag) 2001. Aimed primarily at physicists; provides a self-contained introduction to general relativity, telling you as much as you need to know for cosmological purposes.

Rowan-Robinson, M., *Cosmology*, third edition (Oxford: Oxford University Press) 1996. Has a slightly greater emphasis on astronomical observations than most cosmology texts.

van den Bergh, S., *The Galaxies of the Local Group* (Cambridge: Cambridge University Press) 2000. Cosmology begins at home: contains information about the distance scale within the Local Group, and its dark matter content.

ADVANCED

Kolb, E. W., and Turner, M., *The Early Universe* (Redwood City, CA: Addison-Wesley) 1990. A text that helped to define the field currently known as "particle astrophysics."

Liddle, A. R., and Lyth, D. H., *Cosmological Inflation and Large-Scale Structure* (Cambridge: Cambridge University Press) 2000. Gives an in-depth treatment of inflation, and how it gives rise to structure in the universe.

Peebles, P. J. E., *Principles of Physical Cosmology* (Princeton: Princeton University Press) 1993. A classic comprehensive book.

KEEPING UP TO DATE

Popular astronomy magazines such as *Sky and Telescope* and *Astronomy* provide nontechnical news updates on advances in cosmology. *Scientific American*, from time to time, includes more in-depth articles on cosmological topics. The *Annual Review of Astronomy and Astrophysics* regularly provides reviews, on a more technical level, of cosmological topics.

Table of Useful Constants

	Fundamental Constants
gravitational constant	$G = 6.673 \times 10^{-11} \text{m}^3\text{kg}^{-1}\text{s}^{-2}$
speed of light	$c = 2.998 \times 10^8 \text{m s}^{-1}$
reduced Planck constant	$\hbar = 1.055 \times 10^{-34} \text{J s} = 6.582 \times 10^{-16} \text{eV s}$
Boltzmann constant	$k = 1.381 \times 10^{-23} \text{J K}^{-1} = 8.617 \times 10^{-5} \text{eV K}^{-1}$
electron rest energy	$m_e c^2 = 0.5110 \text{ MeV}$
proton rest energy	$m_p c^2 = 938.272 \text{ MeV}$
neutron rest energy	$m_n c^2 = 939.566 \text{ MeV}$

	Planck Units
Planck length	$\ell_P = (G\hbar/c^3)^{1/2} = 1.616 \times 10^{-35}$ m
Planck mass	$M_P = (\hbar c/G)^{1/2} = 2.177 \times 10^{-8}$ kg
Planck time	$t_P = (G\hbar/c^5)^{1/2} = 5.391 \times 10^{-44}$ s
Planck energy	$E_P = (\hbar c^5/G)^{1/2} = 1.956 \times 10^9 \text{ J} = 1.221 \times 10^{28}$ eV
Planck temperature	$T_P = E_P/k = 1.417 \times 10^{32}$ K

	Conversion of Units
astronomical unit	$1 \text{ AU} = 1.496 \times 10^{11}$ m
megaparsec	$1 \text{ Mpc} = 3.086 \times 10^{22}$ m
solar mass	$1 \text{ M}_\odot = 1.989 \times 10^{30}$ kg
solar luminosity	$1 \text{ L}_\odot = 3.846 \times 10^{26} \text{ J s}^{-1}$
gigayear	$1 \text{ Gyr} = 3.156 \times 10^{16}$ s
electron volt	$1 \text{ eV} = 1.602 \times 10^{-19}$ J

	Cosmological Parameters
Hubble constant	$H_0 = 70 \pm 7 \text{ km s}^{-1}\text{Mpc}^{-1}$
Hubble time	$H_0^{-1} = (4.4 \pm 0.4) \times 10^{17} \text{ s} = 14.0 \pm 1.4$ Gyr
Hubble distance	$c/H_0 = (1.32 \pm 0.13) \times 10^{26} \text{ m} = 4300 \pm 400$ Mpc
critical energy density	$\varepsilon_{c,0} = 5200 \pm 1000 \text{ MeV m}^{-3}$
critical mass density	$\rho_{c,0} = \varepsilon_{c,0}/c^2 = (9.2 \pm 1.8) \times 10^{-27} \text{ kg m}^{-3}$

Index